The Fourier Transform and Its Applications to Optics

The Fourier Transform and Its Applications to Optics

SECOND EDITION

P. M. DUFFIEUX

A Wiley-Interscience Publication

John Wiley & Sons
New York / Chichester / Brisbane / Toronto / Singapore

Originally published as L'intégrale de Fourier et ses applications à l'optique
© Masson, Editeur, Paris, 1970.

Library of Congress Cataloging in Publication Data:

Duffieux, P. M., 1891–
    The Fourier transform and its applications to optics.

    (Wiley series in pure and applied optics, ISSN 0277-2493)
    "A Wiley-Interscience publication."
    Translation of: L'intégrale de Fourier et ses applications à l'optique.
    Includes bibliographical references and index.
    1. Fourier transform optics.   I. Title.   II. Series.

QC381.D8313   1982   535   82-20302
ISBN 0-471-09589-3

Printed in the United States of America

10   9   8   7   6   5   4   3   2   1

Translator's Preface

Duffieux wrote this book in the early forties, and it represents the first book-length treatment of what is now often called Fourier optics.

The contents and presentation of the book, seen in the light of today's knowledge, are remarkable for such an early work. Duffieux's ideas on the decomposition of an object into a spatial frequency spectrum and on diffraction by limited pupils and by gratings, as well as his introduction into optics of the concepts of space invariance and convolution and of the coherent and incoherent transfer functions, were accepted slowly by the optics community. As late as 1955, one of the chief experts on optics wrote: One should not indulge too much in dealing with coherent and semicoherent illumination, frequency analysis, and all straightforward translations of well-known radio communication results into optical technology... the eye is not the ear, and any interpretation of vision in terms of frequencies is really too farfetched. [G. T. di Francia, *Optica Acta* 2, 51, 1955] It was only in the late fifties and early sixties that Duffieux's ideas came into their own, largely as a result of the transfer of ideas from communications engineering.

It will be the task of historians to determine why Duffieux's ideas percolated so slowly into the mainstream of optics. Not having at his disposal the powerful tools of sampling theory, random variables, and distribution theory, Duffieux resorted to a geometrical representation for complex variables, with its attendant vocabulary, that today seems outdated.

The reader will find many early explanations and mathematical derivations. He or she will also find some ideas that have not yet been fully explored, such as the use of the Laue sphere of extension in diffraction theory instead of the Green's function approach, and the concept of the points of information of an image. But for the most part, one will find many of the modern ideas of Fourier optics in their original form.

The book may not appear suitable as a stand-alone text for a course in modern Fourier optics. It should, however, prove useful not only to optics specialists in search of a different perspective on Fourier optics, but also to those interested in the history of optics.

Preface

The second edition of this book, bearing the same title as the first edition, is also designed as a textbook, and it has benefited from the strange manner in which the Fourier transform has evolved since the first edition.

It has become geometric in its most general form. The optical double-diffraction experiment shows that:

1. The focused transform is a distribution of light with the same form as the object;
2. it is directional and spatially bound. This makes it well adapted to the particle model.

Euclid, Descartes, Newton, and Leibnitz developed mathematical theories of shapes that can deal with infinitely small and infinitely large quantities. The part of the world that may be treated by Fourier analysis is the part that is accessible to the senses of man; any object having an extension in space and time is a distribution which may be treated by Fourier analysis. Long ago I considered the example of the elephant Jumbo inside the cell Δx, Δy, Δz.

The successful representation of so-called distributions by means of real or imaginary periodic functions has some strange aspects, and seems to be related in an essential way to the structure of our organs of perception. In 1966, as an introduction to particle optics, I introduced a classification of physical phenomena suggested by the Fourier transform; this classification is reprinted at the end of this introduction. It is valid only at the human level, from statistics to the organization of individuals. The thermal-motional disorder is represented by periodic distributions of photons and phonons.

But the scientific laws of the real world represent phenomena perceived by our senses, and we now know that this knowledge is incomplete and full of traps.

We can see photons, the simplest of all bosons, but we cannot see neutrinos, which come at us along the same paths at the same speed. Photons have allowed us to develop a magnificent theory of wave optics which is an almost perfect mathematical model, but we are as yet unable to manipulate the enormous flux of neutrinos from inside the sun and from terrestrial nuclear reactions. There is a long way to go before physics leads to a complete understanding of our world.

I take the position that the calculations of distributions follow the laws of wave optics, and that the two reciprocal Fourier transforms $F(\cdots)$ and $f(\cdots)$ are continuous. The first five chapters consider the most general Fourier transform with no limitations on the variables. At the end of Chapter 5, Dirichlet's theorem introduces the question of the limits on the extension of the real frequency domain.

Today's students are much more familiar with complex numbers and their meaning than were the students of twenty-five years ago. I have therefore introduced coherent imaging early in Chapter 2. Incoherent imaging then becomes a particular case where the light distribution functions are real and positive. I have also introduced in this edition certain calculations that had appeared, thanks to Aime Cotton, in the *Annales de Physique*. This should make the book more homogeneous and will save the reader from looking up difficult-to-find journals.

From Chapter 6 on, the fundamental Fraunhofer diffraction equation introduces the essential conditions that optics imposes on the Fourier transform:

1. Diffraction is a three-dimensional phenomenon that ties a plane-wave distribution $F(x, y, 0)$ to the propagation of light towards infinity ($z = +\infty$). The resulting light distribution $f(u, v, w)$ may be examined at the focal plane of a lens; u, v, w are the direction cosines of the propagation, obtained by normalizing the wavelength λ to one.

2. We find the Laue sphere for the diffraction of x-rays by a three-dimensional obstacle. The projection of the spherical spectrum on the uv plane is the plane-wave spectrum $f(u, v)$.

3. The spectrum is bounded within the equatorial circle of this sphere, which has a diameter equal to λ, or more generally, restricted to a portion φ of the circle. Dirichlet's theorem clearly indicates that the function $F(x, y, 0)$ is redundant and is composed not of a set of independent points, but of a convolution of itself with the diffraction pattern $\Phi(x, y, 0)$ of the function $\varphi(u, v)$ having range φ and a value of one:

$$F(x, y, 0) = F(x, y, 0) \otimes \Phi(x, y, 0).$$

The function $F(x, y, 0)$ is the Fourier transform of its bounded plane spectrum $f(u, v)$. It is therefore unbounded. If the Fourier integral is replaced by a Fourier series, the result is a discrete function $F_0(x, y)$:

$$F(x, y, 0) = F_0(x, y) \otimes \Phi(x, y, 0).$$

The function $F_0(x, y)$ is not redundant, and it may be bounded. It is the only function with a wavelike origin that can satisfy the principle of conservation of energy, that is bounded, and whose Fourier transform is the limited function $f(u, v)$:

$$f(u, v) = \begin{cases} T[F_0(x, y)] & \text{(series)}, \\ T[F(x, y, 0)] & \text{(integral)}. \end{cases}$$

These two equations are a photon analog of the Maxwell thermokinetic equations, which establish a relation between the distribution of a statistical quantity and its expression at the level of the members of the set under consideration.

The function $f(u, v)$ is then a continuous probability distribution, and $F_0(x, y)$ is a strange complex of photons, affected by a complicated set of interference fringes derived from an analog calculation somewhat similar to an old idea of Newton's. The Fourier transform thus opens a new path towards particle optics.

Those who find it difficult to accept the substitution of the discrete function $F_0(x, y)$ for the wave function $F(x, y)$ may be helped by the example of musical notation, for example the score of the concerto for flute that I am listening to on the radio as I write. It is a Fourier transform: a musical partition where the distance along the staff shows the time and where the traditional frequency scale represents the logarithmic tempered chromatic scale of Bach.

I thank all those who have helped in various ways in the writing of this book.

First of all, the people at the Besançon Optical Laboratory: Professor Vienot, Mrs. Reine the cleaning lady, technicians, machinists, beginning students, and new employees. They have surrounded me with an atmosphere of regular work and of calm that I needed. They have listened to my meanderings, and this has allowed me to write this book: in science one may only write well of that which he has spoken.

I owe special gratitude to:

Christiane Briot, now a teacher, who drew almost all the figures.

Alain Lacourt, a graduate student who read the proofs of my typewritten manuscript.

Jacques Clerc, my office neighbor, who has been my memory for eight years, and who has annotated with red ink my mistakes, my lapses of memory, and my obscure passages. He is the first regular reader for whom this second edition has been written.

Explanation of the table. This is a double entry table. In the column on the extreme left, in descending order of magnitude, are the "worlds" of objects, but especially of concepts, from the cosmos of Alexander Humboldt to the geometrical space-time point with coordinates x, y, z, t. Numerical scales are not given, for they would give an impression of continuity that I wish to avoid. After all, a rat and an elephant are both mammals, and I once sat on a single crystal of white calcite, unfortunately cleaved by the frost, which was at least half a cubic meter in size, and which had been a gigantic molecule.

Each of those "worlds" is built up of a complex of elements from the smaller worlds, and finally from complexes of the molecular, corpuscular, and geometrical worlds. With the exception of the corpuscular world, at least until today, all the objects of the superior worlds are ensembles of homogeneous or disparate elements: all types are possible and acceptable, if some unit may be attributed to them.

When I wanted to fill in this primary classification with some examples, I had too many to choose from. I had to try another classification, not according to our modes of perception, but according to the way things are: things and phenomena.

The best criterion is the degree of correlation between the elements grouped together. We shall see later how to define and measure it. But with fatal uncertainties. The word correlation should be taken in the primitive metaphysical meaning that it had in the medieval translations of Aristotle, and that is still its fundamental meaning in dictionaries of the western world. This concept of correlation represents a loss of independence and of individuality by an element, and also the extent, the complexity, and the force of the links external to the element, but internal to the ensemble; it constitutes the new properties due to the ensemble.

Let us remain within mathematical logic. This degree of logic may go from zero to saturation: the history of a photon may range from free propagation to complete absorption. In the second part of the table, correlation is zero at the left and increases towards the right. At the extreme right are therefore found the individuals, the organized beings. Correlations are essential to their existence, and outside the individual they lose all the meaning that they have within.

P. M. Duffieux

OBJECTS OF THE PHYSICAL SCIENCES

WORLDS	ENSEMBLES Statistics	Correlations	INDIVIDUALS Organization
Cosmic world		– – – – – UNIVERSE – – – – – – – – – – – Diffuse matter Galaxies Stars	
Human world	Radiation Light	– – Tools and mechanisms – – – Gases Liquids Solutions Plasmas Crystals Solids Heat Electricity Magnetism Radioactivity Electronics	Societies Human beings ⋮ Living beings Cells Viruses
Molecular world	Radiation		Large molecules Small molecules Atoms Nuclei
Corpuscular world	Radiation		Corpuscles Photons
Geometrical world	Point (x, y, z, t)	– – – – – – – – – ENERGY – – – – – – – – – – – –	Shapes

Contents

Chapter 1. Fourier Series and Integrals   **1**

Introduction   1
    Distribution over a finite range   1
    Series representation of a limited function   2
    Extension to periodic functions   4
    Simple relations for Fourier series   6
    Symmetry   7
    Alternating functions   9
    Frequency and angular frequency   10
    Combination of odd and even terms   11
    Graphical representation   12
    Terminology   12
  The Fourier integral   13
    Combination of the two integrals   14
    Graphical representation   14
  Relations between series and integrals   15
    Limited isolated function   15
    Isolated functions and associated periodic functions   16
  Change of origin   18
  Uniqueness of the series representation and the Fourier integral   19

Chapter 2. The Frequency Space. The Fourier Transform   **20**

Introduction   20
  Integrals   20
  Series   21

Graphical representation   21
The function $f(u)$ 22
Complex representation   23
Complex functions   25
Complex series and integrals   26
  Complex series   26
  Complex integrals   26
General properties   27
  Additivity   27
  Symmetry and parity   28
The transformation in half-spaces   32
  The real and even transform   32
  The real and odd transform   33
The complex transform   35
Another form of the Fourier transform   35

Chapter 3.   Particular Cases   **36**

Introduction   36
Rectangle function   36
The associated periodic function: the grating   37
The linear ramp   39
Transmittance of a plane diaphragm   40
Associated periodic functions   41
The isosceles triangle   42
Transforms of exponential spectra   43
The Gaussian function   46
The associated periodic function   47
Hermite functions   48
Bessel functions   49
Singular functions   51
General comments   52
  Singular points and jumps   54
  Integrals   55

Chapter 4.   Parseval's Theorem and Convolution   **56**

Introduction   56
Isoplanatic images   56
  Two-dimensional plane images   56
  One-dimensional images   58

Methods for integrating the images   59
Parseval's theorem-convolution   60
  Convolution   61
  Commutativity   61
  Multiple convolutions   62
  Limited functions. Range of the image   62
  Series   63
The point source and the delta function   65
  Physical integrable points   66
  Integrable points and the Fourier transform   67
  General comment   68

Chapter 5.   Applications of Convolution. Dirichlet's Theorem   70

Pinhole camera   70
The scanning slit   71
Resonance curve and impulse response   73
Dirichlet's theorem—weighting functions   75
  Extension of the theorem   77
  Weighting functions   77
  Functions of one variable   80
  Multiple convolutions   83

Chapter 6.   Fraunhoefer Diffraction   86

Introduction   86
  Diffraction of a one-dimensional plane wave   88
  Pupils with two variables   89
  Reciprocity and inverse diffraction   91
  Infinity   92
The physical meaning of the Fourier transform   95
  The refractive index   95
  Abbe's condition   96
  X-ray diffraction   101
  Two-dimensional distributions   104
  The two Fourier transforms   106
  Lambert's law   108
  Range and internal correlation functions for one,
    two, and three variables   108

Chapter 7.   Plancherel's Theorem and Correlation   **112**

Introduction   112
The first form of Parseval's theorem   112
    Product of two functions   112
    The square of a function   113
Plancherel's theorem   114
Application to Fourier series   115
The conservation of energy   116
Graphical representations   116
    Decompositions of complex integrals   116
    Volume of integration of quadratic spectra   119
    Volume of integration of squared functions   120
    Positions of the spectrum of squared functions
        and the quadratic spectrum   123

Chapter 8.   Stigmatic Pupils   **128**

Introduction   128
Interference fringes   128
    Interference by translation   128
    Fresnel's mirrors   131
    On a remark by Wood   132
    A note on symmetrical pupils   136
Stigmatic pupils   137
    The slit pupil   138
Two-dimensional pupils   141
    First procedure   141
    Second procedure   142
    Quadratic spectrum of a square pupil   143
    Quadratic spectrum of a circular pupil   145
    Apodizing pupils   150

Chapter 9.   Discrete Functions   **151**

Introduction   151
The converse of Dirichlet's theorem   152
    Functions of one variable   153
Resolution of a set of points along a line   155
    Bounded sets   156

Incoherent illumination   156
Coherent illumination   157
Information points   158
General remarks   158
Limited discrete pupils   161
Two-dimensional pupils   162
Circular pupils   163
Limited periodic functions—gratings   165
Series of δ-functions   165
Mathematical construction of a finite grating   167
Starting point   167
First operation   168
Second operation   168
Third operation   171
Grating defects   171
The perfect grating   172
The real grating   172
On the definition of Frequency   174
Note on the retinal image   175

Chapter 10.   Transmission of Spatial Frequencies   **176**

Introduction   176
Domains and range functions   176
$\varphi(u, v)$ and $\varphi'(u, v)$ 177
$\varphi(u)$ 178
The sphere of extension and the sphere of radiation   179
Transmission of frequencies   181
Relays and pupil optics   183
Examples   183
Coherent imaging   187
Image of a coherently illuminated slit   188
Abbe's experiment   191
Oblique illumination   193

Index   **195**

The Fourier Transform and Its
Applications to Optics

1

Fourier Series
and Integrals

INTRODUCTION

This chapter contains only the more elementary theorems on Fourier series
and integrals. We have put the emphasis on graphical representations, so
useful to the physicist, and on the relationships between series and integrals,
because those relationships are useful for numerical calculations and for
information theory.

On the other hand we have avoided without regret all existence and
convergence theorems, whose development we fear might bore the reader
impatient to use his mathematics. In general the physicist is interested in
series not just if they are convergent, but only if they are very convergent.
We shall show that in optics all series are finite; the question of their
convergence is therefore irrelevant. Physical laws even suggest that any
conceivable object that can yield an image may always be represented by a
series or by a simple or multiple Fourier integral, even if this series or
integral does not converge and is only a formal representation having
meaning only in certain calculations. We believe that our applications of
those theorems to questions of classical optics is in itself, at least provisionally, a most useful proof.

Distribution over a Finite Range

The functions that the physicist wishes to represent by a Fourier series or
integral usually satisfy *a priori* certain physical conditions which justify the
operation, that is, which lead to convergent series or integrals. These
functions belong to one of the two following classes:

1. The function represents a quantity whose value is defined without
ambiguity for each value of the variable, that always remains finite, and that
may be expressed by a number. This is the case for curves that represent
electrical phenomena as functions of time (voltage, intensity) or meteoro-

logical phenomena (pressure, temperature) or mechanical quantities (speed, elongation), and so on. The function is continuous and finite or measurable. It is only studied or defined over a limited range of the variable, often because it is periodic, but more generally because the phenomenon, or at least its observation, has a finite extent or duration. Such functions are said to have a finite range; for brevity we shall often call them *limited* functions.

2.   The function represents the distribution of something finite: it is a density. This density is again defined without ambiguity at each point of the range of the variable; it is therefore integrable. Thus it is no longer necessary to limit the extent of the function: it may be, or at least be imagined to be, infinite. Thus in optics, where we must represent objects, their images, and illuminated surfaces, the function of interest is illuminance or intensity. In all such cases the distribution of energy, of the vibration amplitudes, or of the momenta, will remain finite.

In practice the only problems that such functions may present are cusps (Fig. 38) or discontinuities (Figs. 31 and 35). For the physicist such a discontinuity of the function or its derivative raises no ambiguity in the definition of the function. The point where the rapid change takes place and that is represented as a discontinuity constitutes a very small interval, of negligible extent. We say in optics that it is not resolvable, and the two values given to the function, or to its derivative, at this point are in reality not simultaneous but successive: before and after the discontinuity. This is the deep meaning of Dirichlet's theorem, which we shall study at length later on.

At the beginning we shall apply the Fourier theory only to functions having such physical meanings. We shall see later, with examples taken from optics, that the theory may be extended to other more complicated cases.

Series Representation of a Limited Function

Consider a uniform and finite function $F(x)$ defined over a range

$$p = x_1 - x_0, \tag{1}$$

of its variable x between x_0 and x_1. This function may be represented at any point over its range by a series of circular functions

$$F(x) = \frac{a_0}{2} + \sum_{n=1}^{\infty} a_n \cos 2\pi n \frac{x}{p} + b_n \sin 2\pi n \frac{x}{p}, \tag{2}$$

where n is an integer.

The feasibility of this representation may be seen through the calculation of the coefficients a_0, $b_0, \ldots, a_n$, $b_n, \ldots$. Let us recall a few definite integrals. Let $t_1 - t_0 = 2\pi$; then if n is an integer,

$$\int_{t_0}^{t_1} \cos nt \, dt = \int_{t_0}^{t_1} \sin nt \, dt = 0,$$

$$\int_{t_0}^{t_1} \cos^2 nt \, dt = \int_{t_0}^{t_1} \sin^2 nt \, dt = \pi, \tag{3}$$

and if m is not equal to n, but both are integers, then

$$\int_{t_0}^{t_1} \cos nt \cos mt \, dt = 0,$$

$$\int_{t_0}^{t_1} \sin nt \sin mt \, dt = 0. \tag{4}$$

But

$$\int_{t_0}^{t_1} \cos nt \sin mt \, dt = 0, \tag{4a}$$

where m and n are any integers.

To calculate the coefficients a_n and b_n, we successively multiply both sides of Eq. (2) by $dx, \ldots, \cos 2\pi n(x/p) \, dx, \sin 2\pi n(x/p) \, dx, \ldots$, and we integrate over the whole range of p for which $F(x)$ is defined. On the right-hand side all the terms except one yield integrals equal to zero. Finally we identify

$$\int_{x_0}^{x_1} F(x) \, dx = a_0 \frac{p}{2}, \tag{5}$$

and

$$\int_{x_0}^{x_1} F(x) \cos 2\pi n \frac{x}{p} \, dx = a_n \frac{p}{2},$$

$$\int_{x_0}^{x_1} F(x) \sin 2\pi n \frac{x}{p} \, dx = b_n \frac{p}{2}. \tag{6}$$

The left sides of the two equations (6) are integrable if $|F(x)|$ is itself integrable. We shall see later that the word integrable must be taken in its wider sense: that of addition. The functions under the integral sign are then said to be integrable in the Lebesgue sense.

To integrate is to sum, to add. A capacitor or a pot of water that is being heated is an integrator. One who throws grains of lead in the pan of a balance integrates by weighing the total quantity of lead on the balance. In Fig. 60 the reader will find a representation of this quantity as a function of time. Very special conventions will be required to represent its derivative. Strictly speaking, this type of function has no derivative; in this type of integration the integrand has little meaning. However, the Fourier transform makes things somewhat easier.

Now this operation of integration, when applied to a differentiated function, leads to a representation of the function, if the range of integration is considered as a variable. Conversely, the knowledge of such a representation allows the immediate evaluation of any integral of the function over a finite interval.

It is regrettable for pedagogical purposes that the two operations are usually designated by the terms "definite integral" and "indefinite integral." Indeed, here the integrals are always definite integrals, in conformity with the strict meaning of the word "integration."

Extension to Periodic Functions

Let the variable k take on all positive and negative integer values. Shifts equal to kp along the x-axis will transform a limited function $F(x)$ into an unbounded periodic function $F(x)$ with infinite range and period p, as illustrated in Fig. 1.

If the range of the variable x in Eq. (2) is infinite, this series represents a periodic function $F'(x)$ from $-\infty$ to $+\infty$, with a period p, because

$$\frac{\cos}{\sin}2\pi n\frac{x+kp}{p}=\frac{\cos}{\sin}2\pi n\frac{x}{p}. \tag{8}$$

This easy and almost automatic extension of the series expression in Eq. (2) often leads to the illusion that the Fourier series necessarily represents periodic functions. Optics quickly dispels this illusion. The coefficients a

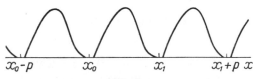

$$x_0-p \qquad x_0 \qquad x_1 \qquad x_1+p \quad x$$

Fig. 1.

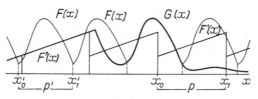

Fig. 2.

and b are always calculated inside a single period. In physics there is no absolutely periodic function; even the Earth's rotation is not perfectly periodic. Within the realm of mathematics, this subtle point is illustrated by the two following cases.

1. *Fragmented function.* Let $G(x)$ be a function defined over any range P. A fragment of $G(x)$ contained in a smaller range p may be represented by a Fourier series S_1. Another fragment contained within another range p' is represented by another series S_2. There is no *a priori* relationship between S_1 and S_2. One may even choose $G(x)$ in such a way that over certain ranges, it is not integrable and may not be represented by a Fourier series. Now, if for each of the two series S the range of the variable x is extended to infinity, the function $G(x)$ is replaced by the new functions $F(x)$ and $F'(x)$ resulting from the infinite repetition of the chosen fragments of $G(x)$; this is shown in Fig. 2.

2. *Isolated limited function.* Consider again the integrable function $F(x)$ defined over a range p, illustrated in Fig. 3. We assume that the function is not repeated outside this range. Along the x-axis, there are an infinity of ranges p' which contain the range p. Outside the range p, the definition of $F'(x)$ is completed by

$$F'(x) = 0 \qquad \text{for} \quad x_0' < x < x_0 \text{ and } x_1 < x < x_1'. \qquad (9)$$

To each new range p' there corresponds a new series S', and obviously to each new range p' there corresponds a new function $F'(x)$, because our

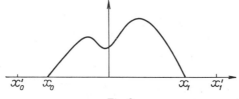

Fig. 3.

original definition must be modified. Now if p' becomes very large, then F' appears as a narrow pulse, and the terms that control its shape move towards the high frequencies. But we shall see that the Fourier integral establishes very simple relations between all the series S' that may be represented by the same curve.

Simple Relations for Fourier Series

There are more ways than one to represent the same function over the same range p:

1.   *General form for any* x_0, x_1.   This is the representation that we have used so far:

$$F(x) = \frac{a_0}{2} + \sum_{n=1}^{\infty} a_n \cos 2\pi n \frac{x}{p} + b_n \sin 2\pi n \frac{x}{p};$$

(10)

$$a_0 = \frac{2}{p} \int_p F(x)\, dx, \qquad \frac{a_n}{b_n} = \frac{2}{p} \int_p F(x) \frac{\cos}{\sin} 2\pi n \frac{x}{p}\, dx.$$

2.   *Normalized form.*   If a_0 is not equal to zero, it is sometimes useful to factor it out, which is equivalent to setting $a_0 = 1$:

$$F(x) = \frac{1}{2} + \sum_{n=1}^{\infty} a_n \cos 2\pi n \frac{x}{p} + b_n \sin 2\pi n \frac{x}{p};$$

(11)

$$1 = \frac{2}{p} \int_p F(x)\, dx, \qquad \frac{a_n}{b_n} = \frac{\int_p F(x) \frac{\cos}{\sin} 2\pi n \frac{x}{p}}{\int_p F(x)\, dx}.$$

3.   *Period* 2π.   The choice of this period often simplifies the equations and the calculations:

$$F(x) = \frac{a_0}{2} + \sum_{n=1}^{\infty} a_n \cos nx + b_n \sin nx,$$

(12)

$$a_0 = \frac{1}{\pi} \int_{2\pi} F(x)\, dx, \qquad \frac{a_n}{b_n} = \int_{2\pi} F(x) \frac{\cos}{\sin} nx\, dx.$$

4. *Origin centered on the range p.* Let $p = 2X$. Then

$$F(x) = \frac{a_0}{2} + \sum_{n=1}^{\infty} a_n \cos \pi n \frac{x}{X} + b_n \sin \pi n \frac{x}{X} ;$$

(13)

$$a_0 = \frac{1}{X} \int_{-x}^{+x} F(x) \, dx, \qquad \frac{a_n}{b_n} = \frac{1}{X} \int_{-x}^{+x} F(x) \frac{\cos}{\sin} \pi n \frac{x}{X} dx.$$

5. *Origin at the beginning of the range p.* We mark the coefficients with primes:

$$F(x) = \frac{a_0'}{2} + \sum_{n=1}^{\infty} a_n' \cos \pi n \frac{x}{X} + b_n' \sin \pi n \frac{x}{X} ;$$

(14)

$$a_0' = \frac{1}{X} \int_0^p F(x) \, dx, \qquad \frac{a_n'}{b_n'} = \frac{1}{X} \int_0^p F(x) \frac{\cos}{\sin} \pi n \frac{x}{X} .$$

There is a simple relationship between the coefficients of Eqs. (13) and (14): the even terms are equal and have the same sign, and the odd terms are equal and have opposite signs:

$$a_0 = a_0',$$

$$a_1 = -a_1', \qquad b_1 = -b_1',$$

$$a_2 = a_2', \qquad b_2 = b_2',$$

(15)

$$a_3 = -a_3', \qquad b_3 = -b_3',$$

$$a_4 = a_4', \qquad b_4 = b_4'.$$

Symmetry

Symmetry may often be used to simplify the series, or at least to simplify the calculations.

1. *Even Functions (Fig. 4).* Let the origin of x be centered on the range p. If the ordinate is an axis of symmetry for a function $F_e(x)$, we have the relation

$$F_e(-x) = F_e(x).$$

(16)

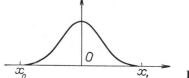

Fig. 4.

All the b coefficients are equal to zero, because the corresponding integrands in (13) are odd on account of the sine:

$$0 = b_1 = b_2 = b_3 = b_4 = \cdots = b_n = \cdots . \tag{17}$$

The series is reduced to the terms containing the a's:

$$F_e(x) = \frac{a_0}{2} + \sum_{n=1}^{+\infty} a_n \cos 2\pi n \frac{x}{p}. \tag{18}$$

The function F_e is said to be even, and the cosine terms are called even terms. They are sometimes called the a-terms.

2. *Odd functions* (*Fig. 5*). If the origin of coordinates is a center of symmetry for $F_o(x)$, so that

$$F_o(-x) = -F_o(x). \tag{19}$$

then the constant term and all the a-terms are equal to zero:

$$F_o(x) = \sum_{n=1}^{\infty} b_n \sin 2\pi n \frac{x}{p}. \tag{20}$$

The function $F_o(x)$ is said to be odd, and all the terms containing sines are called odd terms. They are sometimes called b-terms.

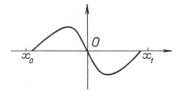

Fig. 5.

In general, any function $F(x)$ may be considered as the sum of an even function F_e and an odd function F_o:

$$F(x) = F_e(x) + F_o(x). \tag{21}$$

From Eqs. (16) and (19), we have

$$F_e(x) = \tfrac{1}{2}[F(x) + F(-x)], \tag{22}$$

$$F_o(x) = \tfrac{1}{2}[F(x) - F(-x)]. \tag{23}$$

The reader may verify that moving the origin to one of the ends of the range p, when the curve which represents the function remains the same, does not change the parity of the series (18) and (20). Only the conditions (16) and (19) change. But this invariance exists only for functions with a finite range. Certain periodic functions may be decomposed into either even or odd components.

Alternating Functions

A function is said to be alternating (Fig. 6) when

$$F\left(x + \frac{p}{2}\right) = -F(x). \tag{24}$$

The constant term and all the terms of even order are equal to zero: only the terms of odd order remain:

$$a_{2k} = b_{2k} = 0 \tag{25}$$

when k is a positive integer or zero. Nevertheless, F is not an odd function.

On the other hand, if a function contains only terms of even order, identical forms will be replicated in the two halves of the range p, as

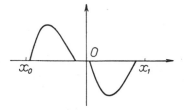

Fig. 6.

illustrated in Fig. 7. The fundamental period is $p/2$:

$$F\left(x + \frac{p}{2}\right) = F(x). \tag{26}$$

Frequency and Angular Frequency

When $F(x)$ is real and infinitely periodic, the nth term is a periodic function having a period p/n. It is useful to consider *frequencies*, which are the inverses of those periods:

$$U = \frac{1}{p}, \qquad u = \frac{n}{p} = nU, \tag{27}$$

where n is equal to a positive integer or zero. Because the order of the term clearly indicates the value of u, we write

$$F(x) = \frac{a_0}{2} + \sum_{n=1}^{\infty} a_n \cos 2\pi ux + b_n \sin 2\pi ux, \tag{28}$$

$$a_0 = 2U \int_p F(x)\, dx, \qquad \frac{a_n}{b_n} = 2U \int_p F(x) \frac{\cos}{\sin} 2\pi ux\, dx. \tag{29}$$

Also useful are the angular frequencies Ω, ω:

$$\Omega = 2\pi U, \qquad \omega = 2\pi u, \tag{30}$$

in terms of which

$$F(x) = \frac{a_0}{2} + \sum_{n=1}^{\infty} a_n \cos \omega x + b_n \sin \omega x; \tag{31}$$

$$a_0 = \frac{\Omega}{\pi} \int_p F(x)\, dx, \qquad \frac{a_n}{b_n} = \frac{\Omega}{\pi} \int_p F(x) \frac{\cos}{\sin} \omega x\, dx. \tag{32}$$

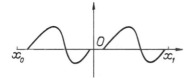

Fig. 7.

According to tradition, the constant term a_0, having a frequency of zero, is always considered separately from the other terms. This distinction is not necessary, because the equations for the a_n and the b_n yield the correct result when n is equal to zero. Indeed, $\cos 0$ is equal to unity and $\sin 0$ is equal to zero. Equation (32) then yields the relation for a_0; the term b_0 thus introduced, like $\sin 0$, is identically zero. The general relation (10) is then written

$$F(x) = \frac{a_0}{2} + \sum_{n=1}^{\infty} a_n \cos 2\pi ux + b_n \sin 2\pi ux,$$

$$\frac{a_n}{b_n} = 2U \int_p F(x) \frac{\cos}{\sin} 2\pi ux \, dx,$$

(33)

where $u = nU$. The two equations may be combined into one:

$$F(x) = \frac{a_0}{2} + 2U \sum_{n=1}^{\infty} \left(\cos 2\pi ux \int_p F(x) \cos 2\pi ux \, dx \right.$$

$$\left. + \sin 2\pi ux \int_p F(x) \sin 2\pi ux \, dx \right), \qquad u = nU.$$

(34)

Combination of Odd and Even Terms

Let

$$a_n = \mathcal{C}_n \cos \Phi_n, \quad b_n = \mathcal{C}_n \sin \Phi_n, \qquad \mathcal{C}_n \geqslant 0. \tag{35}$$

The two terms of order n may be combined. For simplicity, consider the series in the form of Eq. (12) with $p = 2\pi$:

$$a_n \cos nx + b_n \sin nx = \mathcal{C}_n \cos(nx - \Phi_n), \tag{36}$$

$$F(x) = \frac{a_0}{2} + \sum_{n=1}^{\infty} \mathcal{C}_n \cos(nx - \Phi_n) \tag{37}$$

with

$$\mathcal{C}_n^2 = a_n^2 + b_n^2 \quad \text{and} \quad \tan \Phi_n = \frac{b_n}{a_n}. \tag{38}$$

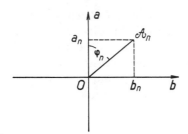

Fig. 8.

Graphical Representation

Following Fresnel, optical scientists have classically represented a harmonic function such as that on the right side of Eq. (36) by a vector of length $\mathcal{C}_n$ rotated by an angle Φ relative to the fixed axis. This is illustrated in Fig. 8. The projections of the vector $\mathcal{C}_n$ on the cartesian axes Oa and Ob are respectively a_n and b_n. It is useful to draw all the vectors in a series on one three-dimensional graph as in Fig. 9. Two of the three axes correspond to a and b; the, third Ou, is the frequency axis, and contains the origins of the vectors $\mathcal{C}_n$. These origins are equidistant with interval U. For a given range p, the frequency axis may therefore be marked off in orders n.

The series of variable terms may be completed by locating a vector equal to $\mathcal{C}_0$ in the plane ab, at a frequency of zero.

Terminology

Graphs such as that of Fig. 9 and Fig. 10 (below) are sometimes called the *spectra* of the functions they represent. In the following chapter we shall

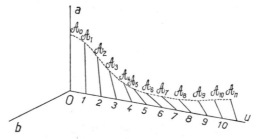

Fig. 9.

redefine the connection between spectrum and function, but the word is often useful and we shall use it here.

THE FOURIER INTEGRAL

Consider an integrable, limited function on the interval p. Let p increase without limit in the expression (34) for the Fourier series. The fundamental frequency U becomes infinitely small; we denote it by du in order to take it inside the summation sign. The discrete sum, in the limit, becomes an integral as the sum becomes continuous. At the same time, the term a_0 disappears. The sum is

$$F(x) = 2\int_0^{+\infty}\left(\cos 2\pi ux \, du \int_{-\infty}^{+\infty} F(x)\cos 2\pi ux \, dx\right.$$

$$\left. + \sin 2\pi ux \, du \int_{-\infty}^{+\infty} F(x)\sin 2\pi ux \, dx\right). \tag{39}$$

The integral on the right side of the equation is the *Fourier integral*. It may be rewritten in a form similar to the series (33):

$$F(x) = 2\int_0^{+\infty}\left[A(u)\cos 2\pi ux + B(u)\sin 2\pi ux\right] dx, \tag{40}$$

$$\begin{matrix} A(u) \\ B(u) \end{matrix} = \int_{-\infty}^{+\infty} F(x) \begin{matrix} \cos \\ \sin \end{matrix} 2\pi ux \, dx. \tag{41}$$

For these equations to be meaningful, it is sufficient that the function $F(x)$ be uniform and integrable on the infinite range of x. It does not matter whether or not the function is limited to the finite range p. It may not, however, be replicated by translation, or superimposed upon itself. If we wish to establish a parallel between Fourier series and integrals, we must set aside periodic functions.

$A(u)$ and $B(u)$ are defined for all values of u between 0 and $+\infty$. They are continuous functions of the variable u. They are also called even and odd components:

1. If $F(x)$ is even, $B(u)$ is identically equal to zero.
2. If $F(x)$ is odd, $A(u)$ is identically equal to zero.

Combination of the Two Integrals

Proceeding in the same way as we have done previously for the series, let

$$A(u) = \mathcal{Q}(u)\cos \Phi(u), \quad B(u) = \mathcal{Q}(u)\sin \Phi(u) \quad \text{with} \quad \mathcal{Q}(u) \geq 0.$$

$$(42)$$

The integral (40) may be written

$$F(x) = 2\int_0^{+\infty} \mathcal{Q}(u)\cos[2\pi ux - \Phi(u)] \, du. \qquad (43)$$

$\mathcal{Q}$, which is always positive, is the modulus, and Φ is the phase, sometimes called the argument. We have

$$\mathcal{Q}(\mathring{u})^2 = A(u)^2 + B(u)^2, \qquad \tan \Phi(u) = \frac{B(u)}{A(u)}. \qquad (44)$$

Graphical Representation

If the range p corresponding to the function of Fig. 9 is allowed to increase without limit, the distance U between the terms tends to zero and the discontinuous ensemble of vectors tends to a continuous surface everywhere perpendicular to plane Oab. Illustrated in Fig. 10 are the two functions $A(u)$ and $B(u)$, which are contained respectively in planes aOu and bOu, and which are the projections of a three-dimensional curve which I shall call, for future reference, $f(u)$. The vector joining the u-axis to a point on the curve has a length equal to the modulus of (43) and an angle equal to its phase.

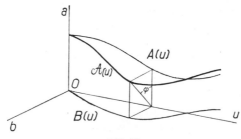

Fig. 10.

RELATIONS BETWEEN SERIES AND INTEGRALS

Limited Isolated Function

Such a function may be represented either as a series over its limited range p, or as an integral over the infinite range of the variable. Consider the series in the form of Eq. (29):

$$\frac{a_n}{b_n} = 2U \int_p F(x) \frac{\cos}{\sin} 2\pi ux \, dx. \qquad (45)$$

For the integral, we have instead

$$\frac{A(u)}{B(u)} = \int_{-\infty}^{+\infty} F(x) \frac{\cos}{\sin} 2\pi ux \, dx. \qquad (46)$$

Now $F(x)$ is zero outside the range p:

$$\int_{-\infty}^{+\infty} = \int_{-\infty}^{x_0} + \int_p + \int_{x_1}^{+\infty} = \int_p. \qquad (47)$$

This means that

$$\frac{a_n}{b_n} = 2U \frac{A(nU)}{B(nU)}, \qquad (48)$$

$$\mathfrak{C}_n = 2U \mathfrak{C}(nU),$$

with the same phase $\Phi_n = \Phi(nU)$.

Except for a factor equal to $2U$, the terms of the series development are the coordinates a and b of the continuous curve $f(u)$. The discontinuous spectrum of $F(x)$ corresponding to the limited range p is found by sampling

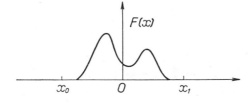

Fig. 11.

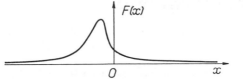

Fig. 12.

the continuous spectrum corresponding to the infinite range of the variable x. The function $f(u)$ therefore represents all the possible series that may represent $F(x)$ over any finite range larger than p. Note that p may not be arbitrarily small, and that the equal distances U between the series terms may not be greater than a certain limit. On the other hand, the distance U may be made as small as desired.

Isolated Functions and Associated Periodic Functions

Consider an isolated function F (Fig. 12) over a finite or infinite range, which is representable by a Fourier transform. Let F be replicated indefinitely with period p along the x-axis. Superpositions, as in Fig. 13, are allowed, and a periodic function is formed by summing at each point along the x-axis the values of all the replications.

Because the function $F(x)$ is real and integrable, the superposition is also real and integrable over any finite range, and the periodic function thus associated with this function may be represented by a Fourier series.

The periodic function is

$$G(x) = \sum_{k=-\infty}^{+\infty} F(x + kp).$$ (49)

The series representing $G(x)$ within the period p is given by

$$\frac{a_n}{b_n} = \frac{2}{p} \sum_{k=-\infty}^{+\infty} F(x + kp) \frac{\cos}{\sin} 2\pi n \frac{x}{p} dx, \qquad k \text{ integral.}$$ (50)

Fig. 13.

The argument may be increased by an arbitrary multiple of 2π:

$$\frac{a_n}{b_n} = \frac{2}{p} \sum_{k=-\infty}^{+\infty} F(x + kp) \frac{\cos}{\sin} 2\pi n \frac{x + np}{p} dx, \tag{51}$$

$$\frac{a_n}{b_n} = \frac{2}{p} \sum_{k=-\infty}^{+\infty} F(x) \frac{\cos}{\sin} 2\pi n \frac{x}{p} dx. \tag{52}$$

This yields Eq. (48):

$$\frac{a_n}{b_n} = 2U \frac{A(nU)}{B(nU)}. \tag{53}$$

The spectrum of the associated periodic function $G(x)$ still results from sampling the continuous spectrum of the solitary function $F(x)$. We shall therefore say that $F(x)$ and $G(x)$ are a function and its associated periodic function, whether there is superposition or not. In Fig. 9 the spectrum of the series and the associated periodic function are represented, except for a factor $2U$, by the vectors and the curve through their tips.

To each isolated function there correspond an infinite number of associated periodic functions, derived from the same spectral curve, but with different periods and a different factor $2U$. Conversely, if the spectrum of a discontinuous or limited function is known, it is always possible to define a limited function associated with a continuous spectrum, by running a continuous, integrable and uniform curve through the tips of the vectors of the discrete spectrum, as shown in Fig. 14. There are ambiguities, but, as we shall see later, they are not always of interest to the optical scientist.

But from the same spectrum of vectors, from a mathematical point of view, it is possible to find an infinite number of associated functions $P(x)$. In Fig. 14 the spectra are real and therefore in a plane, but in general, the

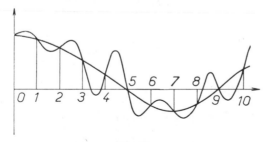

Fig. 14.

spectrum space is three-dimensional and the spectra may be rotated. Each of the functions $F(x)$ thus defined likewise has an infinite number of associated periodic functions.

This association of limited and periodic functions is often useful in numerical analysis and in graphical and mechanical harmonic analysis. In practice, the range over which x may vary is always limited, either by the nature of the phenomena being studied, or because infinity is really too far away. If knowledge is restricted to the range p, the associated periodic functions contain no frequencies smaller than $U = 1/p$.

In Chapter 3, we shall see a number of examples of such associations.

CHANGE OF ORIGIN

From the preceding considerations, it is sufficient to consider only the case of the isolated function represented by a Fourier integral. Let this function be $F(x)$; after the change of variables

$$x' = x + t, \tag{54}$$

let

$$F'(x) = F(x' - t) = F(x). \tag{55}$$

The two functions $F(x)$ and $F'(x)$ represent the same curve, the same "object," except for a shift t along the x-axis. The development of $F'(x)$ into a Fourier integral is given by

$$\begin{matrix} A'(u) \\ B'(u) \end{matrix} = \int_{-\infty}^{+\infty} F(x) \begin{matrix} \cos \\ \sin \end{matrix} 2\pi u(x + t) \, dx, \tag{56}$$

$$\begin{matrix} A'(u) \\ B'(u) \end{matrix} = \cos 2\pi ut \int_{-\infty}^{+\infty} F(x) \begin{matrix} \cos \\ \sin \end{matrix} 2\pi ux \, dx$$

$$\mp \sin 2\pi ut \int_{-\infty}^{+\infty} F(x) \begin{matrix} \sin \\ \cos \end{matrix} 2\pi ux \, dx, \tag{57}$$

$$\begin{matrix} A'(u) \\ B'(u) \end{matrix} = \cos 2\pi ut \begin{matrix} A(u) - \\ B(u) + \end{matrix} \sin 2\pi ut \begin{matrix} B(u) \\ A(u) \end{matrix}. \tag{58}$$

From the last two equations

$$\mathcal{Q}'(u) = \mathcal{Q}(u),$$

$$\Phi'(u) = \Phi(u) + 2\pi ut. \tag{59}$$

For a given frequency, the modulus of the new function is the same, but the phase is increased by an angle proportional to t and to u. The curve of the spectrum $f'(u)$ is therefore derived from $f(u)$ by a rotation of the vectors $\mathcal{Q}(u)$ about the u-axis, by an amount proportional to the abscissa u and to the coefficient t.

We shall call this deformation of the curve $f(u)$ a helical twist. The spectra of two functions that differ only by a translation along the x-axis have graphs or curves that differ only by a helical twist about the axis. For us two such functions are considered to be identical.

UNIQUENESS OF THE SERIES REPRESENTATION AND THE FOURIER INTEGRAL

For a given function $F(x)$ and a given range of integration, the expressions that we have found may lead to a series or Fourier integral representation. It may be shown that this representation is unique, and that any other method of calculation would lead to the same result. It may also be shown that two different functions may not be represented by the same Fourier series or integral. For those questions, the reader is referred to more detailed mathematical works.

2

The Frequency Space.
The Fourier Transform

INTRODUCTION

The lack of symmetry between the x-space extending between $-\infty$ and $+\infty$ and the exclusively positive u-space makes Fourier series and integrals almost useless for optical imagery. Fortunately, it is easy to introduce negative frequencies into the series and integrals of the previous chapter.

The graphical representations and the difficulties associated with trigonometric equations point immediately towards the use of complex exponentials. At the price of a small effort, extending complex numbers to functions and frequencies leads to simpler and more compact equations, easier to remember and to manipulate. We shall also see how this widening of the mathematical domain leads to a widening of the optical domain.

Integrals

Consider again Eq. (1.39):

$$F(x) = 2\int_0^{+\infty} \left(\cos 2\pi ux \, du \int_{-\infty}^{+\infty} F(x)\cos 2\pi ux \, dx \right.$$

$$\left. + \sin 2\pi ux \, du \int_{-\infty}^{+\infty} F(x)\sin 2\pi ux \, dx \right). \qquad (1.29)$$

If the variable u is changed to $-u$ in the integrated functions, the results of the integrations are not changed, because the signs of the cosines do not change, and the two sines both change sign, so the sign of their product remains the same. If the range of the first integration is extended from $-\infty$ to $+\infty$, the quantity under the integral sign is multiplied by two, and the equality may be reestablished by suppressing the factor 2 in the initial expression. This leads immediately to the expressions corresponding to Eqs.

(1.39), (1.40), and (1.41):

$$F(x) = \int_{-\infty}^{+\infty}\left(\cos 2\pi ux\,du \int_{-\infty}^{+\infty} F(x)\cos 2\pi ux\,dx\right.$$

$$\left. + \sin 2\pi ux\,du \int_{-\infty}^{+\infty} F(x)\sin 2\pi ux\,dx\right), \tag{1}$$

$$\begin{matrix} A(u) \\ B(u) \end{matrix} = \int_{-\infty}^{+\infty} F(x)\begin{matrix}\cos\\\sin\end{matrix} 2\pi ux\,dx, \tag{2}$$

$$F(x) = \int_{-\infty}^{+\infty}[A(u)\cos 2\pi ux + B(u)\sin 2\pi ux]\,du. \tag{3}$$

Series

The expressions (1.33) and (1.34) are for series what Eqs. (1.40) and (1.41) are for integrals. The extension of n to negative values changes nothing in the integrations, except for suppressing the factor 2 in the expressions for $F(x)$, a_n, and b_n. The frequencies here are multiples of a fundamental frequency, which, in order to mark the analogy, we denote by

$$U = \frac{1}{p}. \tag{4}$$

This yields a group of three equations:

$$F(x) = U \sum_{n=-\infty}^{+\infty}\left(\cos 2\pi nUx \int_p F(x)\cos 2\pi nUx\,dx\right.$$

$$\left. + \sin 2\pi nUx \int_p F(x)\sin 2\pi nUx\,dx\right), \tag{5}$$

$$\begin{matrix} a_n \\ b_n \end{matrix} = U \int_p F(x)\begin{matrix}\cos\\\sin\end{matrix} 2\pi nUx\,dx, \tag{6}$$

$$F(x) = \sum_{n=-\infty}^{+\infty} a_n\cos 2\pi nUx + b_n\sin 2\pi nUx. \tag{7}$$

Graphical Representation

Graphical representations are derived from Fig. 9 for series and from Fig. 10 for integrals in the same way as before. It is sufficient to note that the

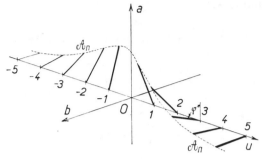

Fig. 15.

coefficients a and A belong to even functions and the coefficients b and B to odd ones:

$$a_{-n} = a_n, \qquad A(-u) = A(u),$$
$$b_{-n} = -b_n, \qquad B(-u) = -B(u). \tag{8}$$

The negative parts of the new graphs and the positive parts already determined are symmetrical with respect to the a-axis, as illustrated in Figs. 15 and 16.

The Function $f(u)$

It is easy to see that $F(x)$ may be written in the form of Eq. (1.43):

$$F(x) = \int_{-\infty}^{+\infty} \mathcal{C}(u)\cos[2\pi ux - \Phi(u)]\, du. \tag{9}$$

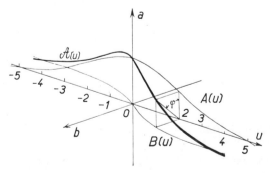

Fig. 16.

It is obvious that the helix in Fig. 16 constructed from the two projections $A(u)$ and $B(u)$ represents a function which we have already called $f(u)$, and which may be expressed in a simple way only after we have accepted complex functions of the real variables x and u.

We shall see in the last chapters of this book that all the preceding considerations may be applied to functions of three variables x, y, z in a cartesian space, with the corresponding frequencies u, v, w. The functions $F(x, y, z)$ and $f(u, v, w)$ are in two complex spaces and are related by a Fourier transform. One might think of the space u, v, w as an abstract space; in fact it may be observed as easily as the Cartesian space x, y, z. Optical instruments make one space just as tangible as the other. To distinguish them, we call the x, y, z space the real space $\mathcal{R}$, and the u, v, w space the Fourier or frequency space $\mathcal{F}$.

COMPLEX REPRESENTATION

A complex variable, which is the sum of a real and an imaginary part

$$z = a + ib, \tag{10}$$

is a useful artifice whenever a vector length ρ is related to an angle φ. Let

$$|z| = \rho = +\sqrt{a^2 + b^2},$$
$$\tan \varphi = \frac{b}{a}; \tag{11}$$

the variable z may be written in the well-known form

$$z = \rho(\cos \varphi + i \sin \varphi) = \rho \operatorname{cis} \varphi, \tag{12}$$

$$z = \rho e^{i\varphi} = \rho \exp(i\varphi), \tag{13}$$

where ρ is the modulus, which is always positive, and φ is the argument, or in the language of optics, the phase.

The complex number z may be represented by a vector in the rectangular coordinate system with axes a and b, where b is the imaginary axis, and a is the real axis. This is illustrated in Fig. 17.

Let us recall a few elementary operations.

The sum of two complex numbers is equal to the sum of the vectors that represent them, as illustrated in Fig. 18. The resultant is equal to z, where

$$z' = a' + ib', \qquad z'' = a'' + ib''; \tag{14}$$

$$z = z' + z'' = (a' + a'') + i(b' + b''). \tag{15}$$

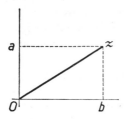

Fig. 17.

The multiplication of two complex quantities may be considered as a formal operation on exponentials:

$$z' = \rho' \text{cis}\, \varphi', \qquad z'' = \rho'' \text{cis}\, \varphi'', \tag{16}$$

$$z = z' + z'' = \rho' \rho'' \text{cis}(\varphi' + \varphi''). \tag{17}$$

The modulus of the product is the product of the moduli, and the argument is the sum of the arguments.

Two complex quantities are said to be *conjugate* when their arguments are equal but of opposite signs. For a given complex quantity z, the complex conjugate is found by replacing $+i$ by $-i$ everywhere. The symbol for conjugation is an asterisk:

$$z = a + ib = \rho \, \text{cis}\, \varphi, \qquad z^* = a - ib = \rho \, \text{cis}(-\varphi). \tag{18}$$

The sum of two quantities that are complex conjugates of each other is real:

$$z + z^* = 2a = 2\rho \cos \varphi. \tag{19}$$

The product of two complex conjugates is real and equal to the square of their common modulus:

$$zz^* = \rho^2 \text{cis}(\varphi - \varphi) = \rho^2. \tag{20}$$

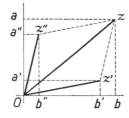

Fig. 18.

COMPLEX FUNCTIONS

A function is said to be complex when it may be decomposed into the sum of a real and of an imaginary part:

$$F(x) = F'(x) + iF''(x), \qquad (21)$$

where $F'(x)$ and $F''(x)$ are two real functions. $F(x)$ may then be represented as a left-handed helix in the system of coordinates

$$\begin{aligned} &x, \\ a = {}&F'(x), \qquad\qquad (22) \\ b = {}&F''(x). \end{aligned}$$

The two functions F' and F'' are the projections of $F(x)$ on the planes xOa and xOb, which are called the real and the imaginary planes, and which are illustrated in Fig. 19.

$P(x)$ and $\Phi(x)$ are the modulus and the phase of $F(x)$:

$$|F(x)| = P(x) = +\sqrt{F'(x)^2 + F''(x)^2}, \qquad (23)$$

$$\tan \Phi(x) = \frac{F''(x)}{F'(x)}. \qquad (24)$$

$F(x)$ may be written as a complex exponential, which is more suggestive and easier to manipulate. The geometrical meaning of $P(x)$ and $\Phi(x)$ is immediate from Fig. 19.

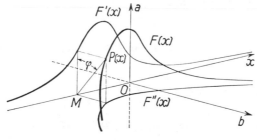

Fig. 19.

It is obvious that $f(u)$ has the same form as $F(x)$:

$$f(u) = A(u) + iB(u) \tag{25}$$

$$= \rho(u)\text{cis } \varphi(u), \tag{26}$$

$$F(x) = P(x)\text{cis } \Phi(x). \tag{27}$$

COMPLEX SERIES AND INTEGRALS

Complex Series

We shall follow step by step the calculations made for real series.

Let $F(x)$ be an absolutely integrable function defined over the range p; it may be represented by the complex series

$$F(x) = \sum_{n=-\infty}^{+\infty} Z_n\text{cis}(2\pi nUx), \tag{28}$$

where $U = 1/p$ is any positive, negative, or zero integer, and the coefficients Z are complex. It is sufficient to show that it is possible to calculate Z.

Multiply the two sides of Eq. (28) by $\text{cis}(-2\pi mUx) \, dx$, where m is any integer, and integrate over the range p. For any $m \neq n$, the integral on the right is equal to zero, but for $m = n$ the exponential on the right side reduces to unity and the integral on the right side is equal to $Z_n p$. Equation (28) becomes

$$Z_n = U\int_p F(x)\text{cis}(-2\pi nUx) \, dx. \tag{29}$$

Complex Integrals

Complex integrals follow from complex series in the same way that real integrals follow from real series. Equations (28) and (29) may be combined to yield

$$F(x) = \sum_{n=-\infty}^{+\infty} \text{cis}(2\pi nUx) \cdot U\int_p F(x)\text{cis}(-2\pi nUx) \, dx. \tag{30}$$

Now substitute $nU = u$ and $U = du$:

$$F(x) = \sum_{-\infty}^{+\infty} \text{cis}(2\pi ux) \, du \int_p F(x)\text{cis}(-2\pi ux) \, dx. \tag{31}$$

Assuming that the integrals remain finite, if p is increased without limit, the sum may be replaced by an integral:

$$F(x) = \int_{-\infty}^{+\infty} \text{cis}(2\pi ux)\, du \int_{-\infty}^{+\infty} F(x)\text{cis}(-2\pi ux)\, dx. \qquad (32)$$

Now let

$$f(u) = \int_{-\infty}^{+\infty} F(x)\text{cis}(-2\pi ux)\, dx, \qquad (33)$$

$$F(x) = \int_{-\infty}^{+\infty} f(u)\text{cis}(2\pi ux)\, du. \qquad (34)$$

These two equations, similar except for the sign of i, are the reciprocal equations of the complex Fourier transform. They link the two reciprocal spaces $\mathcal{R}$ and $\mathcal{F}$. They show that the two conjugate variables x and u behave as frequencies with respect to each other. It remains only to analyze the properties of this transformation and to draw the consequences for optics.

GENERAL PROPERTIES

We shall consider the best-known and simplest case: the reciprocal complex Fourier transform that we have just defined. The right sides of Eqs. (33) and (34) contain two conjugate operators

$$T[\cdots] = \int_{-\infty}^{+\infty} [\cdots]\text{cis}(2\pi ux)\, du, \qquad (35)$$

$${}^*T[\cdots] = \int_{-\infty}^{+\infty} [\cdots]\text{cis}(-2\pi ux)\, dx. \qquad (36)$$

From now on, defining equations will be written in the short form

$$F(x) = T[f(u)],$$
$$f(u) = {}^*T[F(x)]. \qquad (37)$$

Additivity

Certain obvious conclusions may be drawn from the form of the operators:

The Fourier transform of the sum of two functions is equal to the sum of their Fourier transforms:

$$F(x) + G(x) = T[f(u) + g(u)],$$
$$f(u) + g(u) = {}^*T[F(x) + G(x)]. \qquad (38)$$

If a function is multiplied by a factor which is not a function of the transformation variables, its Fourier transform is multiplied by the same factor:

$$KF(x) = KT[f(u)],$$

$$G(y)F(x) = G(y)T[f(u)], \qquad (39)$$

$$(a + ib)F(x) = (a + ib)T[f(u)].$$

Symmetry and Parity

An arbitrary complex function contains a real part $F'(x)$ and an imaginary part $iF''(x)$, where $F''(x)$ is real. Each of those two real functions is the sum of an even part F'_e or F''_e, and an odd part F'_o or F''_o; $F(x)$ may therefore be expanded into four terms,

$$F = (F'_e + F'_o) + i(F''_e + F''_o), \qquad (40)$$

Its transform may also be expanded in the same way:

$$f = (f'_e + f'_o) + i(f''_e + f_o). \qquad (41)$$

The eight functions on the right in the last two expressions are plane functions in the complex space x, a, b, and their planes contain the x-axis. We now generalize: if the plane is arbitrary, it is defined by its angle φ with the real plane aOx. The following lemma holds:

If a plane function is multiplied by $\exp(i\psi)$, *its plane is rotated by an angle ψ about the x-axis.*

In particular if ψ is equal to $\pi/2$, then the factor is

$$\exp(i\pi/2) = i.$$

If the plane function was real, it becomes imaginary; if the plane function was imaginary, it becomes real.

A second lemma may be deduced from Eqs. (2), (3), and (4) and from Fig. 16 to establish relations of parity between the eight elements of F and f:

If a real function F' has an even part and an odd part, the Fourier transform of the even part is real and even; the Fourier transform of the odd part is imaginary and odd.

If an imaginary function has an even part and an odd part, the Fourier transform of the even part is imaginary and even, and the Fourier transform of the odd part is real and odd.

Thus the Fourier transform conserves parity.

Because the Fourier transformation is additive, the Fourier transform of Eq. (40) may be written

$$f(u) = (\varphi'_e + i\varphi'_o) + i(\varphi''_e + i\varphi''_o). \tag{42}$$

This establishes an immediate correspondence with the decomposition of Eq. (41):

$$f(u) = (\varphi'_e - \varphi''_o) + i(\varphi''_e + \varphi'_o). \tag{43}$$

At this point let us derive a few results which will prove very useful later. From Eq. (40) the conjugate of F may be obtained be replacing i by $-i$:

$$*F = (F'_e + F'_o) - i(F''_e + F''_o). \tag{44}$$

Let us call the transform of this $f_j(u)$. In Eq. (42), the i inside the parenthesis is changed to $-i$:

$$f_j = (\varphi'_e + i\varphi'_o) - i(\varphi''_e + i\varphi''_o); \tag{45}$$

that is,

$$f_j(u) = (\varphi'_e + \varphi''_o) - i(\varphi''_e - \varphi'_o). \tag{46}$$

When u is changed to $-u$, the signs of the odd functions change:

$$f_j(-u) = (\varphi'_e + \varphi''_o) - i(\varphi''_e + \varphi_o)' = f^*(u). \tag{47}$$

To illustrate this section, we present in Figs. 20–26 a number of cases of symmetry and parity, to which have been added the three following cases:

$F(x)$ is any real function.
$F(x)$ is any even function.
$F(x)$ is any odd function.

Note. All the plane components of the functions $F(x)$ and $f(u)$ will be described and drawn in the plane in Chapter 3. The positive directions for the x-axis and the u-axis are the same, and may be interpreted directly only

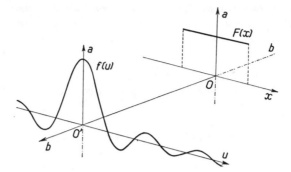

Fig. 20. F and f real and even; $F_e' = T(f_e')$.

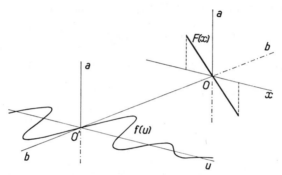

Fig. 21. F real and odd, f imaginary and odd; $F_o' = iT(f_o')$.

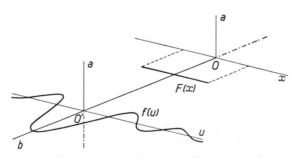

Fig. 22. F and f imaginary; $F_e'' = T(f_e'')$.

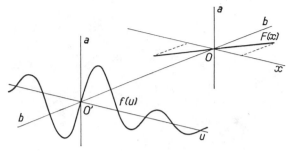

Fig. 23. *F* imaginary and odd, *f* real and odd; $F_o'' = T(f_o'')$.

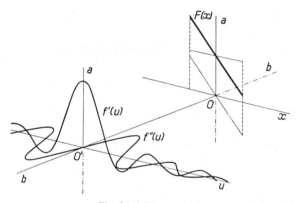

Fig. 24. *F(x)* real.

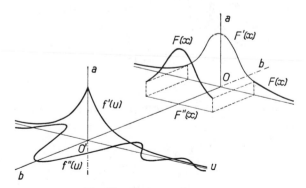

Fig. 25. *F(x)* complex, even.

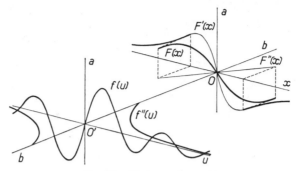

Fig. 26. $F(x)$ complex, odd.

by an observer located between the origins O and O' on the imaginary axis b. This is the disposition that will be adopted for the relation between a pupil $F(x)$ and its Fraunhofer diffraction pattern $f(u)$.

THE TRANSFORMATION IN HALF-SPACES

The following will be done only for integrals: enough has been said about how to go from integrals to series.

The Real and Even Transform

Let $F(x)$ be real and even. From Eq. (3) and Eq. (4),

$$F(x) = \int_{-\infty}^{+\infty} A(u)\cos 2\pi ux \, dy,$$

$$A(u) = \int_{-\infty}^{+\infty} F(x)\cos 2\pi ux \, dx. \tag{48}$$

$A(u)$ and $F(x)$ are a Fourier-transform pair. From our previous considerations on parity

$$F(x) = F'_e(x),$$

$$A(u) = f'_e(u). \tag{49}$$

The parity allows the integration to be limited to the positive values of the variable, so that

$$F(x) = 2\int_0^{+\infty} A(u)\cos 2\pi ux\, du,$$

$$A(u) = 2\int_0^{+\infty} F(x)\cos 2\pi ux\, dx. \tag{50}$$

This is the Fourier-transform pair for real and even functions.

The Real and Odd Transform

Let $F(x)$ be real and odd; again from Eq. (3) and Eq. (4),

$$F(x) = \int_{-\infty}^{+\infty} B(u)\sin 2\pi ux\, du,$$

$$B(u) = \int_{-\infty}^{+\infty} F(x)\sin 2\pi ux\, dx. \tag{51}$$

In the half-spaces Ox and $O'u$, the real and odd transform pair may be written

$$F(x) = 2\int_0^{+\infty} B(u)\sin 2\pi ux\, du,$$

$$B(u) = 2\int_0^{+\infty} F(x)\sin 2\pi ux\, dx. \tag{52}$$

These transforms are often used in electricity and in electronics when $F(x)$ is defined only for positive values of x. This is the case for any phenomenon starting at the time zero. They are also found in the Laplace transform, and in all cases when variables with circular symmetry replace the Cartesian coordinates.

Either of the two transforms is applicable in most cases, because a function may be extended either as an even or as an odd function. However, the physics sometimes imposes a choice.

Consider for example the function

$$F(x) = \begin{cases} 1 & \text{for } 0 < x < X, \\ 0 & \text{for } x < 0 \text{ and } X < x, \end{cases} \tag{53}$$

which may be considered as the positive part of the rectangle function

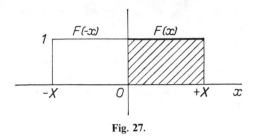

Fig. 27.

illustrated in Fig. 27. Its Fourier integral in the frequency half-space is

$$F(x) = 2 \int_0^\infty \frac{\sin 2\pi uX}{2\pi uX} \cos 2\pi ux \, du, \tag{54}$$

with

$$A(u) = 2 \int_0^X \cos 2\pi ux \, dx. \tag{55}$$

But the same function may also be considered as the positive part of the odd function illustrated in Fig. 28. The corresponding odd Fourier-transform pair is

$$F(x) = 2 \int_0^\infty \frac{1 - \cos 2\pi uX}{2\pi uX} \sin 2\pi ux \, du,$$

$$B(u) = 2 \int_0^X \sin 2\pi ux \, dx. \tag{56}$$

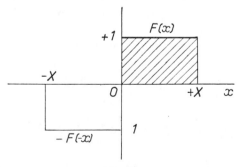

Fig. 28.

The Complex Transform

Let us consider the half-space for both the variables x and u. If we are looking for a complex Fourier transform, we find a double ambiguity: both the real and imaginary parts may be considered as belonging either to even or to odd functions. There are therefore four systems of conjugate equations, which will not be developed here.

But the reader will remember that I concluded the previous chapter with the classical statement on the uniqueness of the Fourier series and integral of a given function. Obviously, this is valid only if the coordinate space and the frequency space are both extended to their whole algebraic range, thus eliminating contradictory interpretations.

Another Form of the Fourier Transform

In the expressions for the Fourier transform used so far, substitute x' for x and u' for u:

$$F(x') = \int_{-\infty}^{+\infty} f(u')\mathrm{cis}(2\pi u'x')\, du',$$

$$f(u') = \int_{-\infty}^{+\infty} F(x')\mathrm{cis}(-2\pi u'x')\, dx'. \tag{57}$$

Now let

$$\sqrt{2\pi}\, u' = u, \qquad du' = \frac{1}{\sqrt{2\pi}}\, du,$$

$$\sqrt{2\pi}\, x' = x, \qquad dx' = \frac{1}{\sqrt{2\pi}}\, dx. \tag{58}$$

When these values are substituted in Eq. (57), a new form for the Fourier transform is obtained:

$$F(x) = \frac{1}{\sqrt{2\pi}} \int_{-\infty}^{+\infty} f(u)\mathrm{cis}(ux)\, du,$$

$$f(u) = \frac{1}{\sqrt{2\pi}} \int_{-\infty}^{+\infty} F(x)\mathrm{cis}(-ux)\, dx. \tag{59}$$

This is the form most frequently used by mathematicians. For certain purposes, such as studying polynomials or Hermite functions, it is more useful than ours. But it is less useful for applications in optics.

3

Particular Cases

INTRODUCTION

Among the following examples, many will be needed later; some are given to demonstrate the power of the representation by Fourier series and integrals; in addition, a few functions have been included for their value as examples, especially in the figures. These are all classical examples, and the reader may skip them if he finds the list too long or too detailed. More complete collections of Fourier-transform pairs have been published.

We have not developed some integrals which are easy, and in any case may be found in Dwight's *Table of Integrals*. All the integrals are calculated and illustrated in the complete spaces $\mathcal{R}$ and $\mathcal{F}$, with the variables x and u extending from $-\infty$ to $+\infty$.

The cases already considered have led to some general considerations about the resources of the Fourier transform. We have based them, without proof, on considerations which will soon be derived for the optical scientist, from the harmonic theory of optical images.

Some of the following examples have already been considered in the previous chapter, to illustrate our discussion on symmetry. We leave their detailed analysis to the reader.

RECTANGLE FUNCTION

Suppose $F(x)$ is equal to a constant A in the range $-X$ to $+X$ (Fig. 29). Let A be real and positive: all other cases may be derived from this one. Then

$$F(x) = \begin{cases} A & \text{for} \quad -X < x < +X, \\ 0 & \text{for} \quad x < -X, \quad +X < x. \end{cases} \qquad (1)$$

Because the function is real and even, its Fourier transform is also real and even:

$$f(u) = 2AX\frac{\sin 2\pi uX}{2\pi uX}. \qquad (2)$$

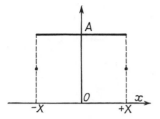

Fig. 29.

So $F(x)$ is equal to

$$F(x) = 2AX\int_{-\infty}^{+\infty} \frac{\sin 2\pi uX}{2\pi uX} \cos 2\pi uX\, du. \qquad (3)$$

$f(u)$ has periodic zeros at the points

$$u = mU \quad \text{with} \quad U = 1/2 \text{ and } m \text{ an integer;}$$

when $m = 0$, $f(u)$ has an absolute minimum equal to $2AX$.

THE ASSOCIATED PERIODIC FUNCTION: THE GRATING

Let the rectangle function be periodically repeated, without superposition or connection with the period

$$p = 2X' > 2X$$

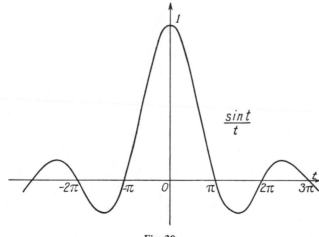

Fig. 30.

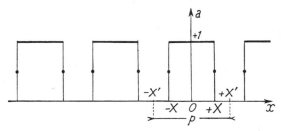

Fig. 31.

(see Fig. 31). This function is still real, but it has periodic gaps of width $2(X' - X)$.

The transform (Fig. 32) is still real and even if $F(x)$ is real and even, that is, if the origin of the variable x is centered on one of the rectangles, or on one of the gaps. Let it be centered on a rectangle. Let

$$U' = \frac{1}{p},$$

$$f(u) = 2AU'X\frac{\sin 2\pi nU'X}{2\pi nU'X} \qquad \text{with } n \text{ an integer,} \qquad (4)$$

$$F(x) = 2AU'X \sum_{n=-\infty}^{+\infty} \frac{\sin 2\pi nU'X}{2\pi nU'X} \cos 2\pi nU'X. \qquad (5)$$

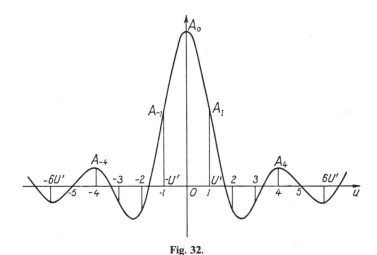

Fig. 32.

The terms A_n of the series may be derived from the previous function $f(u)$; their period is equal to $U' = 1/p$.

If p and $2X$ are commensurable, some of the terms of the series will be equal to zero. Later we shall study the Ronchi grating, which consists of alternating white and black bands of equal width, and for which

$$p = 4X, \qquad U = 2U'.$$

Except for the term A_0, all the even terms coincide with zeros of $f(u)$ and are equal to zero. Only the term A_0 and the odd terms remain.

THE LINEAR RAMP

The linear function

$$y = ax \qquad \text{with} \quad a = \text{constant}$$

may only be represented by a Fourier series if it is limited to a finite range of the variable x. For the simplest case (Fig. 33), let

$$F(x) = \begin{cases} \dfrac{A}{X}x & \text{for} \quad -X < x < +X, \\ 0 & \text{for} \quad x < -X, \quad +X < x. \end{cases} \tag{6}$$

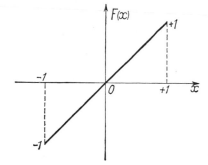

Fig. 33.

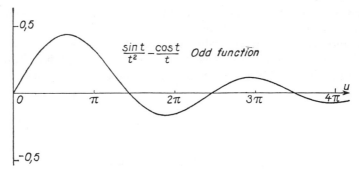

Fig. 34.

The transform $f(u)$ is odd and imaginary. We drop the factor i which, in the case of a plane function, places it in the plane Oxb (Fig. 34):

$$f(u) = 2AX\left[\frac{\sin 2\pi uX}{(2\pi uX)^2} - \frac{\cos 2\pi uX}{2\pi uX}\right].$$ (7)

TRANSMITTANCE OF A PLANE DIAPHRAGM

Let a thin opaque diaphragm be pierced by an aperture of area S. Let i be the angle of incidence of a uniform collimated beam of light. The transmittance at the aperture is equal to $S\cos i$, and is positive whether the incidence is from one side or the other. Let $\cos i = w$. It is easy to see in Fig. 35 that $g(w)$ is the even function which corresponds to the odd function of Fig. 33. Using here the notation that is defined at the end of Chapter 6,

$$g(w) = S|w| \qquad \text{for} \quad -1 < w < +1,$$ (8)

and zero elsewhere.

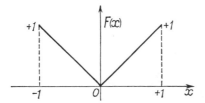

Fig. 35.

The Fourier transform $G(z)$ is also real and even:

$$G(z) = S \left[\frac{\sin 2\pi z}{2\pi z} - \frac{1 - \cos 2\pi z}{(2\pi z)^2} \right]. \qquad (9)$$

Figure 36 illustrates the positive part of this even function. We shall see it again in more detail in Chapter 6.

ASSOCIATED PERIODIC FUNCTIONS

Two such functions will be given. The first, $F'(x)$ (Fig. 37a) is derived directly from the preceding linear function:

$$F'(x) = x \qquad \text{for} \quad -\pi < x < +\pi$$

$$= \sum_{n=-\infty}^{+\infty} (-1)^{n+1} \frac{\sin nx}{n}. \qquad (10a)$$

The second function, illustrated in Fig. 37b, will be useful later:

$$F''(x) = \pi - x \qquad \text{for} \quad 0 < x < 2\pi,$$

$$= \sum_{n=-\infty}^{+\infty} \frac{\sin nx}{n}. \qquad (10b)$$

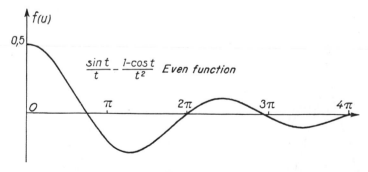

Fig. 36.

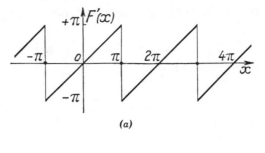

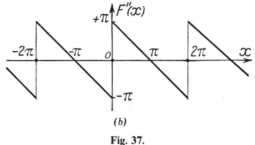

(a)

(b)

Fig. 37.

THE ISOSCELES TRIANGLE

Consider the even function illustrated in Fig. 38 and derived from the linear ramp

$$F(x) = \begin{cases} 0 & \text{for } x < -X, \\ \dfrac{A}{X}(x + X) & \text{for } X < x < 0, \\ \dfrac{A}{X}(X - x) & \text{for } 0 < x < +X, \\ 0 & \text{for } +X < x. \end{cases}$$

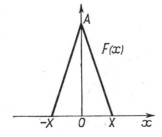

Fig. 38.

Its Fourier transform, a real and even positive definite function shown in Fig. 39, is

$$f(u) = AX\left(\frac{\sin \pi uX}{\pi uX}\right)^2. \tag{11}$$

Note. If the base of the triangle were twice as long, the range of the triangle would extend from $-2X$ to $+2X$, and its Fourier transform would be

$$f(u) = 2AX\left(\frac{\sin 2\pi uX}{2\pi uX}\right)^2. \tag{12}$$

Let us compare this with the rectangle function and its transform, assuming that the values of A and X are the same. The coefficient $2AX$, which is equal to the area of the rectangle and of the new triangle, is the same. The function of u is the same, but it is squared for the triangle. In Chapter 6, we shall see the physical parallel to this mathematical relation.

TRANSFORMS OF EXPONENTIAL SPECTRA

The function e^{-au}, where a is a positive coefficient, is integrable from 0 to $+\infty$, but is not integrable in the half-space of negative frequencies. It is possible to limit $f(u)$ to the positive half-space of frequencies, while remaining in the domain of sine and cosine transforms. The function is shown in Fig. 40.

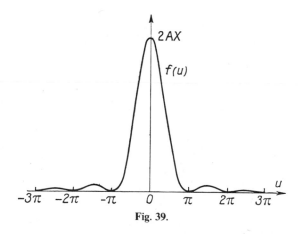

Fig. 39.

1. *The simple even form* (*Fig.* 41).

$$F(x) = 2\int_{0}^{+\infty} e^{-au}\cos 2\pi ux\, du \qquad \text{with} \quad a > 0$$

$$= \frac{2a}{a^2 + 4\pi^2x^2}. \tag{13}$$

2. *The associated even periodic form* (*Fig.* 41). These patterns correspond to the multiple wave interference fringes of Boulouch and Fabry:

$$F(x) = 1 + 2\sum_{n=1}^{+\infty} r^n\cos nx, \qquad n \text{ a positive integer and } 0 < r < 1$$

$$= \frac{1 - r}{1 - 2r\cos x + r^2}. \tag{14}$$

The discrete spectrum is illustrated in Fig. 40.

3. *The odd form.* In Fig. 42 we give without proof the equations and the curves for $F(x)$. The curve for $f(u)$ is the same as for the even forms, except that it is located in the b-plane instead of the a-plane.

Aperiodic function:

$$F(x) = 2\int_{0}^{+\infty} e^{au}\sin 2\pi ux\, du \qquad \text{with} \quad 0 < a$$

$$= \frac{4\pi x}{a^2 + 4\pi^2x^2}. \tag{15}$$

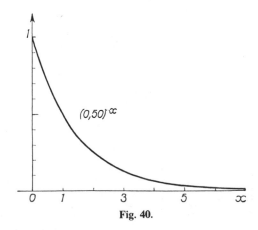

Fig. 40.

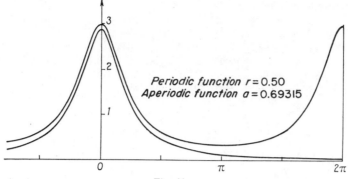

Fig. 41.

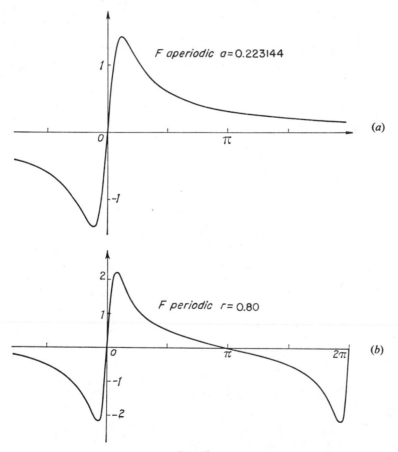

Fig. 42.

Associated periodic function:

$$F(x) = 2 \sum_{n=1}^{+\infty} r^n \sin nx \qquad \text{with} \quad 0 < r < 1$$

$$= \frac{2r \sin x}{1 - 2r \cos x + r^2}. \tag{16}$$

THE GAUSSIAN FUNCTION

This function plays an important role in Fourier theory:

$$f(u) = \exp(-a^2 u^2). \tag{17}$$

The transform of this real and even function is, according to the rules for symmetry, also real and even. But the transform $F(x)$ has the same form as $f(u)$:

$$F(x) = \int_{-\infty}^{+\infty} \exp(-a^2 u^2) \cos(2\pi ux) \, du,$$

$$= \frac{\sqrt{\pi}}{a} \exp\left(-\frac{\pi^2}{a^2} x^2\right) \qquad \text{with} \quad a > 0. \tag{18}$$

Let

$$F(x) = \frac{\sqrt{\pi}}{a} \exp(-b^2 x^2). \tag{19}$$

The relation

$$a^2 b^2 = +\pi^2 \tag{20}$$

shows that a and b are inversely proportional to each other. In particular, if

$$a = b = +\pi, \tag{21}$$

the Fourier-transform pair have identical coefficients and exponents:

$$T\left[\exp(-\pi u^2)\right] = \exp(-\pi x^2),$$

$$T\left[\exp(-\pi x^2)\right] = \exp(-\pi u^2). \tag{22}$$

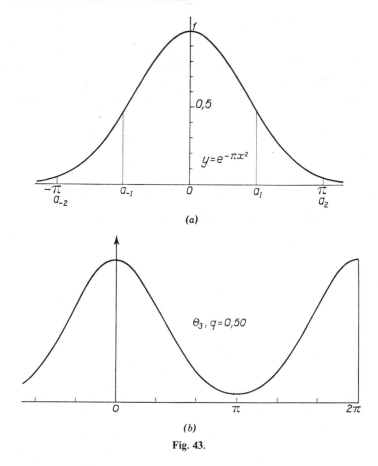

(a)

(b)

Fig. 43.

The well-known bell-shaped graph of the Gaussian function is illustrated in Fig. 43*a*.

THE ASSOCIATED PERIODIC FUNCTION

The associated periodic function is the Theta function represented in Fig. 43*b*:

$$\theta_3\left(\frac{x}{p}\right) = \sum_{n=1}^{+\infty} q^{n^2}\cos\left(2\pi n\frac{x}{p}\right), \text{ with } n \text{ any integer and } 0 < q < 1. \quad (23)$$

The discrete spectrum of this function is given in Fig. 43a; its envelope is the Gaussian function.

HERMITE FUNCTIONS

Hermite functions are a fairly easy extension of the Fourier transform, but not very handy to use. A few of their properties are given without proof.

Let $f(t)$ and $g(s)$ be a Fourier-transform pair in the form of the last paragraph of Chapter 2, which was used by Hermite. There exists an operator that is common to the two functions. It is a simple matter to show that

$$T\left[\frac{d^2f(t)}{dt^2} - t^2f(t)\right] = \frac{d^2g(s)}{ds^2} - s^2g(s). \tag{24}$$

If η is an arbitrary variable, the operator may be written

$$\left(\frac{d^2}{d\eta^2} - \eta^2\right)f(\eta), \tag{25}$$

Equation (24) then becomes

$$\left(\frac{d^2}{dt^2} - t^2\right)f(t) = \left(\frac{d^2}{ds^2} - s^2\right)g(s). \tag{26}$$

For this kind of transformation, it is the function $t^2/2$ that is its own transform:

$$T\left[\exp\left(-\frac{t^2}{2}\right)\right] = \exp\left(-\frac{s^2}{2}\right). \tag{27}$$

Hermite looked for functions

$$\Phi_n(t) = H_n\exp\left(-\frac{t^2}{2}\right), \tag{28}$$

where H_n is a polynomial of degree n which satisfies the differential equation

$$\frac{d^2\Phi}{dt^2} - t^2\Phi = \lambda\phi, \tag{29}$$

in which λ is a factor that depends only on n. H_n must satisfy the differential equation

$$\frac{d^2H}{dt^2} - 2t\frac{dH}{dt} + 2nH = 0, \tag{30}$$

which yields the polynomials H by recurrence. Finally, we have

$$T\left[\Phi_n(t)\right] = (-i)^n\Phi_n(s). \tag{31}$$

From the functions Φ_n are derived normalized and orthogonal functions ψ_n, which yield transformations analogous to the Fourier transform. The first few functions Φ are shown in Fig. 44. The functions of even order are even, and those of odd order are odd. These functions are useful for the calculations of apodized diffraction patterns. It is curious that the function of order $2k$ can give K lateral fringes to the central fringe of the exponential.

BESSEL FUNCTIONS

For all practical purposes, the content of this section consists of its title. There are many Bessel functions and their algebra is complicated, even when limited to what is used in optics for problems involving circular symmetry. For details the reader is referred to specialized works.

The first of the two kinds of Bessel functions, J_p, for which we shall give the transform equations, is very important not only for instrumental optics, but also in other experimental studies on light. The ordinary spectrum of two-dimensional images is bounded by a circle, and its ultimate extension is

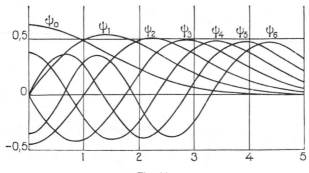

Fig. 44.

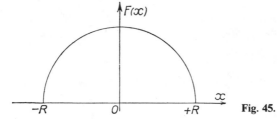

Fig. 45.

a circle of radius $2/\lambda$. We consider only the simple case where $f(u, v)$ is a constant inside such a circle. The Fourier-transform pair $F(x, y)$ and $f(u, v)$ have circular symmetry, and the amplitude distribution along a diameter of F is conjugate to that along a diameter of f:

$$F(x) = \begin{cases} +\sqrt{R^2 - x^2} & \text{for} \quad -R < x < +R, \\ 0 & \text{for} \quad x < -R, +R < x, \end{cases} \tag{32}$$

$$F(x) = \int_{-\infty}^{+\infty} \pi R^2 \frac{J_1(2\pi Ru)}{2\pi Ru} \cos 2\pi ux \, du, \tag{33}$$

$$T[F(x)] = \pi R^2 \frac{J_1(2\pi Ru)}{2\pi Ru}. \tag{34}$$

$F(x)$ may be represented by the semicircle of radius R illustrated in Fig. 45.

The functions $f(u) = J_1(t)/t$ is a cosine transform. Therefore its normal Fourier transform is even.

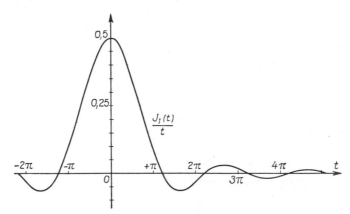

Fig. 46.

SINGULAR FUNCTIONS

1. The Bessel function $J_0(t)$ is an even function. Its transform is another simple function

$$F(x) = \begin{cases} \dfrac{1}{+\sqrt{R^2 - x^2}} & \text{for} \quad -R < x < +R \\ 0 & \text{elsewhere.} \end{cases} \tag{35}$$

$F(x)$ is positive and infinite at the extremities of its range: the integral does not converge.

2. It is possible to integrate the function

$$F(x) = \frac{a^2}{\pi x(a^2 + 4\pi^2 x^2)}. \tag{36}$$

Indeed, we have the integral

$$\int_{-\infty}^{+\infty} F(x)\sin 2\pi ux \, dx = 1 - e^{-au} \quad \text{with} \quad au > 0. \tag{37}$$

On the other hand the integral of the exponential does not converge between 0 and $+\infty$.

The two functions are represented in Fig. 47. This spectrum will be compared later with that of Dirac's δ-function.

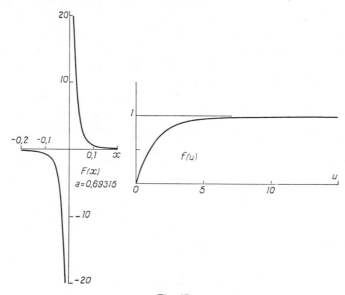

Fig. 47.

GENERAL COMMENTS

The preceding examples show the extraordinary flexibility of the representation by Fourier series and integrals. The functions may have cusps, discontinuities, gaps, and integrable singularities (as will be seen later), but one must never forget to specify the range over which the density of the distribution is equal to zero, and the points at which it becomes infinite while remaining integrable.

For most physicists, the first contact with Fourier representations involves periodic phenomena: the tones of musical instruments, electrical or elastic vibrations, alternating current. That is why Fourier theories are often summarized in a statement like the following:

Any phenomenon with any shape that is periodic with frequency v_0 may always be decomposed into a series of harmonic (or sinusoidal) functions with frequencies that are multiples of the fundamental frequency v_0.

The series is a particular case of the integral, and the periodic function has no special quality to distinguish it as the fundamental kind of function related to Fourier theories. On the contrary, because of its infinite range, it is not integrable. It is subject to Fourier theories only because it may be defined over one of its periods. It is in fact integrable nonperiodic functions that are the proper subject of Fourier theories. Such functions represent something finite over some range, which may be finite or infinite, and it is the repetition of those functions that allows us, through an additional convention, to go to periodical functions. It is this finite character of the information (or the energy) that makes the Fourier integral so well adapted to the representation of optical images.

More accurately, the Fourier series and integrals may be considered as increasingly precise approximations to this finite something. For example, take the series for which the explanations are the simplest and the most immediate.

For the function of Fig. 48, the term $a_0/2$, which represents the average density over the finite range p, represents the totality of this something

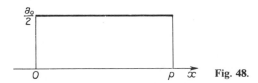

Fig. 48.

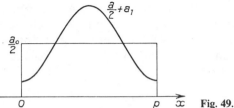

Fig. 49.

spread out uniformly over the range p. The first-order term only recasts the uniform distribution as a sine distribution, as shown in Fig. 49. The mechanism of the successive corrections by the higher-order harmonics is a bit less simple to visualize. It is the same from the second term on, so we shall explain it only for this term. Consider it alone, on the auxiliary axis in Fig. 50. Now by vertically sliding the ordinate as illustrated in Fig. 50, distort it, along with the plotted sine wave, so that it takes the form of the preceding partial sum S_1. This gives us the displacement that must be imposed on the distribution S_1 to give it the form of S_2.

It now becomes apparent that these displacements of our quantity, imposed by successive deformations, become increasingly localized as the order of the sum S_n increases. In fact, the quantity of stuff pushed out by one crest of the sinusoid is exactly compensated by the next trough.

When the reader considers the proof that we propose in Chapter 5 for Dirichlet's theorem, based upon convolution, he will understand why, in the first edition of this book, we rejected the classical proof of this theorem in favor of a version adapted to the needs of optics. When one enters upon the theory of functions following Fourier, he must be resigned to the informa-

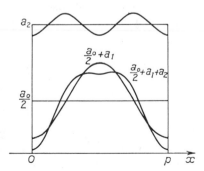

Fig. 50.

tion at one point being a reflection of the information over the whole function, and not of the information from one element progressively and artificially refined about that point. This will not surprise the optical scientist, who knows that one grain of dust on an optical surface may illuminate the whole field of the instrument.

Singular Points and Jumps

It is impossible, for a finite n, to make the curvature of S_n equal to zero except at an inflection point, or to create a discontinuity. It is obviously impossible to make the curvature of S_n zero over a continuous range if n remains finite. From these remarks may be derived the following properties of functions representable by Fourier series or integrals:

1. A series with a finite number of terms may not represent a function with a discontinuous slope.
2. A series with a finite number of terms may not represent a function with a straight-line segment over a finite range.
3. A series with a finite number of terms may not represent a function with a gap.
4. A series with a finite number of terms may not represent a function with a discontinuity.

Functions that have discontinuities in the slope, straight-line segments, jumps, or discontinuities may only be represented by series or integrals that are infinite.

It is easy to imagine how such an infinite series converges to the function that it finally represents. The partial sum S_n consisting of the constant terms and the first n terms represented in Fig. 51 is the oscillating curve; as n increases, the amplitudes of the oscillations become smaller and the sum tends towards the limiting curve.

Inside an infinitely small interval of the variable where the sum of the series has no discontinuity, the amplitudes of the variations of S_n tend to

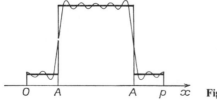

Fig. 51.

zero as n increases indefinitely. Things are different when the interval contains a discontinuity, or is limited by a discontinuity such as that at A; in that case, S_n tends to a particular value given by Dirichlet's classical theorem. The behavior of the partial sums S_n and of the corresponding curves is known in mathematics as the Gibbs phenomenon. We shall see later that it was well known to physicists under the name of the diffraction phenomenon or diffraction fringes.

Integrals

The theorems for series and integrals given in Chapter 1 allow the extension of all the properties that we have just given for series, to integrals that represent isolated functions over the infinite range of the variable.

4

Parseval's Theorem and Convolution

INTRODUCTION

We shall abandon here the rigorous methods of mathematical analysis in favor of the more inductive, more direct, more flexible, more physical, more conventional, and mathematically more dangerous methods of the physicist. Considering our intentions, it seems difficult to explain Parseval's theorem and its consequences in a useful way without invoking both coherent and incoherent exposition.

Optical images are usually two-dimensional, like our immediate perception of things. We shall see in the next chapter that x-ray diffraction, despite the three-dimensional nature of its objects and of its spectra, itself yields only two-dimensional images. This is doubtless because the third dimension is occupied by the propagation, but mostly because our retina and our emulsions record only two-dimensional images.

We shall reduce these two dimensions to one and use functions of one variable as often as possible: their equations take up less space, and their representation in complex space already requires three coordinate axes. There is no loss of generality in using only the pair ux of conjugate variables. This pair may symbolically represent the forms $ux + vy$ and $ux + vy + wz$ of two-dimensional and three-dimensional imaging.

The examples given in Chapter 5 to illustrate the theorems of this chapter are not restricted to the field of optics, because the idea of an image which is the result of convolution is much wider than the field of optics. There is no longer any science that can lock itself behind the walls of a secret garden. We shall have plenty of time later to consider optical images.

ISOPLANATIC IMAGES

Two-Dimensional Plane Images

Consider a plane object that may be described by its luminance distribution $F_0(x, y)$, where (x, y) is a point on the object. In the image plane of this plane object, the image may be completely described by a function that gives the distribution of illuminance over the plane. If the optical instrument were perfect, that is, geometrically stigmatic, it would establish a point-by-point correspondence between the object and the image, each point of the image receiving light only from the conjugate point of the object. In this case, knowledge of the image would be strictly equivalent to knowledge of the object, even if the correspondence law established a deformation—a systematic distortion of the object. All of the information would be transmitted. Let $F(x, y)$ be the distribution of illuminance in this "perfect" image, which we call a *stigmatic* image of the object.

In real optical images, the energy from an infinitesimal point is always dispersed over an extended area of the image. The energy, which in a stigmatic image should remain in an infinitesimal element $dx\, dy$, is spread out over a spot, which is the effective image of the corresponding point of the object. In order to formulate the problem in a simple and usable form, we make two logical assumptions:

1. If the luminance of the object point is multiplied by a factor k, the illuminance of all the points of its effective image are multiplied by this factor. The shape of the spot does not change.

2. If the object point is shifted in its plane, the effective image is shifted in the image plane without changing its shape, its orientation, or its illuminance. This is the definition of isoplanatism.

Let $D(x, y)$ be the observed effective image (Fig. 52). In order to distinguish its coordinates from those of the stigmatic image $F(x, y)$, we shall add primes to the latter, so the stigmatic image is written $F(x', y')$. The axes of x and x' and of y and y' are exactly superimposed.

Let

$$x - x' = \xi, \qquad y - y' = \eta. \tag{1}$$

Consider a real image $F(x, y)$ in terms of energy. The quantity of energy

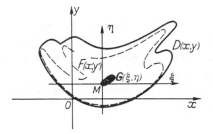

Fig. 52.

available in the element $dx'\,dy'$ of the stigmatic image is equal to $F(x',\,y')\,dx'\,dy'$. From the two preceding assumptions, this energy is distributed according to the function

$$[F(x',\,y')\,dx'\,dy']G(\xi,\,\eta), \qquad (2)$$

where the quantity inside the brackets is only a coefficient, and the function $G(\xi,\,\eta)$ or $G(x - x',\,y - y')$ is independent of x and y.

We call this function G the dissipation figure or spread function. It depends on the effective image and on the optical instrument. Its invariance in the image plane is called the condition of homogeneity or isoplanatism. The effective image is therefore a homogeneous image, or in classical instrumental optics, an isoplanatic image.

If the object is incoherent, the superposition at a point $x',\,y'$ of the energy coming from different elements add like scalar quantities, and the effective image at a fixed point $x,\,y$ is the sum at this point of all that is sent there by the elements of $F(x',\,y')$:

$$D(x,\,y) = \int_{-\infty}^{+\infty}\int_{-\infty}^{+\infty} F(x',\,y')G(x - x',\,y - y')\,dx'\,dy'. \qquad (3)$$

When the image is coherent, it is not a scalar sum but a vector sum of the interference that is produced at the point $x,\,y$. The same equation, with complex functions, yields the effective coherent image.

We shall meet two-dimensional images again, but in order to treat the fundamental equations with a minimum of complications, we shall reduce Eq. (3) to an expression between functions of one variable that we call the *one-dimensional image*.

One-Dimensional Images

The parallel expression to Eq. (3) now contains only one variable:

$$D(x) = \int_{-\infty}^{+\infty} F(x')G(x - x')\,dx'. \qquad (4)$$

We shall consider right off the most general problem to be found in optics, where F, G, and their Fourier transforms are complex functions of two conjugate real variables, x and u. Real functions are a particular case of this, when the imaginary parts are set equal to zero.

Before considering the Fourier transform, two practical methods of integrating $D(x)$ will be given.

Methods for Integrating the Images

In a system with three coordinate axes Ox, Oa, and Ob, where Oa is the real axis and Ob is the imaginary axis, $F(x)$ is represented by a left-handed helix in Figs. 53 and 54. In order to keep the figures simple, only the projection $F''(x)$ of $F(x)$ on the imaginary plane xOb is represented on the figures. The spread function $G(x)$ has been chosen simple enough for the figures to remain comprehensible. It is the plane complex triangle ABC whose base BC coincides with the x-axis and whose apex is in the plane aOb; it presents no symmetry.

For both geometrical interpretations, it is necessary to construct a triangle $A'B'C'$ derived from triangle ABC.

First Procedure (Fig. 53).   Let O' be a point at x' along the x-axis. The vector $O'P$ is the modulus of $F(x)$, and φ is its argument. Let the axes Oa, Ob, and $G(x)$ be duplicated. This ensemble is now given a translation OO' followed by a rotation φ about Ox, bringing the duplicate of Oa into coincidence with $O'P$. The coordinates a and b of $G(x)$ are then multiplied by $O'P$, becoming the coordinates a' and b' of a new curve G', which is the triangle $A'B'C'$ in Fig. 53. The curve G' represents the product $F(x')G(x - x')$, where x is still the variable coordinate.

This product is represented at a point M of the abscissa x, which determines the constant of integration, by the vector MM', and $MM'\,dx'$ is the element of integration. The integration is carried out by sliding the curve G' along the x-axis while making it follow the fluctuations of $O'P$.

This procedure is geometrically complicated, but we shall see that for real functions, it leads to a simple geometrical representation.

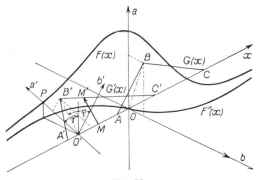

Fig. 53.

Second Procedure (Fig. 54).   This procedure is founded upon the reflection of $G(x)$. Let

$$G'(x) = G(-x). \tag{5}$$

The curve of $G'(x)$ is the mirror image of the curve $G(x)$ in to the complex plane aOb. We again double the axes Oa and Ob; let x', a', and b' be the coordinates of the new triangle. $G'(x)$ becomes $G'(x')$ or $G(-x)$. This ensemble is now shifted by x' along the x-axis, moving the origin from the point O to O', where we want to determine the convolution product. It remains to integrate the product of F and G', which is equal to $F(x')G(x' - x)$, that is, $F(x')G(x - x')$.

For images which are two-dimensional functions of x and y, the second procedure is the most useful. It should also be remembered that convolution is commutative and that the roles of F and G may be interchanged in the preceding constructions.

PARSEVAL'S THEOREM—CONVOLUTION

Let $F(x)$ and $G(x)$ be two real or complex functions whose Fourier transforms are respectively $f(u)$ and $g(u)$. Let $F(x)$ be the stigmatic image, and $G(x)$ the spread function. The image is

$$D(x) = \int_{-\infty}^{+\infty} F(x')G(x - x') \, dx', \tag{6}$$

where x' is the variable of integration, whose scale coincides with that of x.

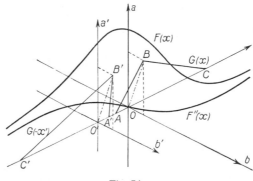

Fig. 54.

Substitute for G its Fourier transform

$$G(x) = \int_{-\infty}^{+\infty} g(u)\text{cis}(2\pi ux)\, du. \tag{7}$$

Because x appears only in the exponential, it may be replaced by $x - x'$:

$$D(x) = \left[\int_{-\infty}^{+\infty} F(x') \int_{-\infty}^{+\infty} g(u)\text{cis}[2\pi u(x - x')]\, du \right] dx'. \tag{8}$$

The integrals are separated by factoring out of the second integral dx' and the exponential containing x':

$$D(x) = \int_{-\infty}^{+\infty} F(x')\text{cis}(2\pi ux')\, dx' \int_{-\infty}^{+\infty} g(u)\text{cis}(2\pi ux)\, du. \tag{9}$$

The first integral is precisely $f(u)$, which, being a function of u, goes under the second integral sign:

$$D(x) = \int_{-\infty}^{+\infty} f(u)g(u)\text{cis}(2\pi ux)\, du. \tag{10}$$

The Fourier transform of $D(x)$ is the product of the Fourier transforms of the two functions $F(x)$ and $G(x)$:

$$d(u) = f(u)g(u). \tag{11}$$

Convolution

The operation which in the language of optics we have called homogeneous or isoplanatic spread is, in the language of mathematics, convolution; in abridged form it may be written

$$D(x) = F(x) \otimes G(x). \tag{12}$$

The second member of this expression is a convolution product. In the Fourier space u there corresponds the algebraic product of Eq. (11).

Commutativity

Because the algebraic product is commutative, so is the convolution product:

$$D(x) = F(x) \otimes G(x) = G(x) \otimes F(x), \tag{13}$$

$$\int_{-\infty}^{+\infty} F(x')G(x - x')\, dx' = \int_{-\infty}^{+\infty} F(x - x')G(x')\, dx'. \tag{14}$$

Multiple Convolutions

In many optical instruments such as periscopes and endoscopes, information is transmitted from the object $F_0(x)$ to a final image $F_n(x)$ through a series of intermediate images $F_1, F_2, \ldots, F_{n-1}$, each corresponding to a different convolution with spread functions $G_1, G_2, \ldots, G_n$:

$$F_0 \otimes G_1 \otimes G \otimes \cdots \otimes G_n = F_n. \tag{15}$$

The relation between the transforms is the product

$$f_0 g_1 g_2 \cdots g_n = f_n. \tag{16}$$

From a mathematical point of view, because convolution is commutative, the order of the convolutions is arbitrary. From a physical point of view this is rarely possible.

The relation between the transforms is interesting from more than one point of view: it yields the ultimate that may be hoped for in transmission. Each of the successive convolutions is a frequency filter, and they may be described by the vocabulary of electrical communications.

The equations give the term-by-term transmission, called the frequency transfer, defining the distortions. If the transfer of the moduli is a function of the frequency, there is amplitude distortion between F_0 and F_n. If the phase change is a function of frequency, there is a phase distortion.

There is one kind of distortion which is never present for isoplanatic spread functions: the appearance of new frequencies whose amplitudes were zero in the original stigmatic image; this is called frequency distortion. This phenomenon is common and even normal in photography.

Limited Functions. Range of the Image

Let two functions $F(x')$ and $G(\xi)$, defined as in the preceding sections, be limited to the finite ranges p and q:

$$X_0 < x' < X_1 \quad \text{and} \quad p = X_1 - X_0,$$

$$\xi_0 < \xi < \xi_1 \quad \text{and} \quad q = \xi_1 - \xi_0. \tag{17}$$

The range of x that contributes to the formation of an element dx of the image $D(x)$ is given by the inequality

$$\xi_0 < x - x' < \xi_1. \tag{18}$$

When Eq. (17) is added to Eq. (18), we get

$$X_0 + \xi_0 < x < X_1 + \xi_1. \tag{19}$$

The range of $D(x)$ becomes

$$(X_1 + \xi_1) - (X_0 + \xi_0) = p + q = P. \tag{20}$$

This is illustrated in Fig. 55.

The image is thus independent, at least in a certain sense, of the position of $G(x)$ relative to $F(x)$, and of its absolute position. This position may be changed by a translation t along x. In the Fourier plane there corresponds a spiral twist $\operatorname{cis}(-2\pi ut)$ of the spectrum $g(u)$:

$$T[G(x + t)] = \operatorname{cis}(-2\pi ut)T[G(x)]. \tag{21}$$

The spiral translation factor multiplies the two sides of Eq. (11). The image is therefore shifted by t without deformation.

Series

The extension of the range of the image outside of the ranges p and q of the functions F and G allows convolutions to be applied to Fourier series. But certain precautions must be taken.

CASE 1.   This is the simplest case: F and G are expanded in a series over a range p' which contains the two ranges p and q, or more simply, the sum $p + q$ of the two ranges:

$$p' > P = p + q. \tag{22}$$

The image $D(x)$ is contained entirely and by itself in each period p'. Figure

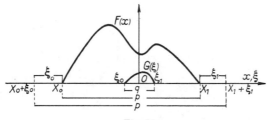

Fig. 55.

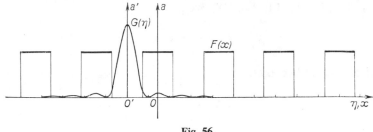

Fig. 56.

55 illustrates this simple case: a sufficient condition is that p' be equal to or greater than P.

CASE 2.   One of the two functions, say $F(x)$, is periodic or has been made periodic with a period p. G is a nonperiodic function, continuous and limited or unlimited. This is the case for the grating shown in Fig. 56, with parallel lines long enough and numerous enough to be treated as infinitely periodic, as well as for its diffraction pattern.

The discrete spectrum of $F(x)$ is multiplied by the continuous spectrum of $G(x)$, as illustrated in Fig. 57. We have represented the spectrum of the non-periodic function $F(x)$ without taking into account the factor $1/p$. Here it is the spread function of $F(x)$ that must be considered as a periodic function, because of its discrete spectrum.

CASE 3.   The two functions are periodic with the same period p, and their two spectra are superimposed exactly with the period $U = 1/p$. To multiply those two spectra together is exactly equivalent to the previous operation, except for the scale factor U. The integration may thus be carried out on only one period, and the convolution is produced over the infinite range of the two periodic functions.

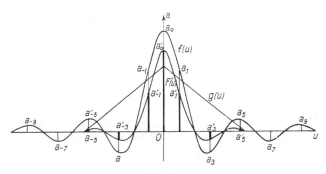

Fig. 57.

If the periodic function $G(x)$ of this third case is considered to be produced by a periodic replication with a period p of the $G(x)$ of the preceding case with the unlimited nonperiodic function, the two cases are seen to be fundamentally the same.

CASE 4. The last case is that where F and G are periodic functions with different periods p and q. The spectral terms only coincide occasionally, if the two periods are commensurable. There can be no convolution analogous to the previous case. Let us put aside this case, which has some relation to moiré patterns.

THE POINT SOURCE AND THE DELTA FUNCTION

The infinitesimal point, located by its coordinates, was given to us by Euclidian geometry, so we call it a Euclidian point. Many quantities have been attached to it by modern science: Newton gave it a mass, astronomers have counted point sources in the sky according to their magnitudes, optical scientists have adopted it, J. J. Thompson gave it an electric charge; particle physics has carried on.

From these concepts Dirac has abstracted his delta function, indicating within parentheses the variable with respect to which it is integrable. If the singularity is at the origin, the function is written $\delta(x)$; if at the abscissa x_0, it is written $\delta(x - x_0)$.

It is easy to introduce the delta function into convolutions and the Fourier transform. The convolution of $F(x)$ with $\delta(x)$ yields

$$F(x) \otimes \delta(x) = F(x) \cdot 1 = F(x). \tag{23}$$

Let $H(\theta)$ be the Fourier transform of $\delta(x)$; then

$$f(u)H(\theta) = f(u), \tag{24}$$

$$H(\theta) = 1 \quad \text{from } -\infty \text{ to } +\infty, \quad \text{(Fig. 58a)}. \tag{25}$$

If $\delta(x)$ is replaced by $\delta(x - x_0)$,

$$F(x) \otimes \delta(x - x_0) = F(x - x_0), \tag{26}$$

the transform $H(x_0)$ represents the helical twist corresponding to the translation x_0:

$$H(x_0) = \exp(-i2\pi u x_0) \quad \text{(Fig. 58b)}. \tag{27}$$

$F(x)$ is a density with magnitude A:

$$F(x) = \frac{dA}{dx}.\tag{28}$$

On the scale of F, δ has an infinite density: $1/0$.

We shall later meet in two and in three dimensions integrable lines and surfaces with similar definitions, and which may be considered to be formed of points or δ-functions.

Physical Integrable Points

For physical integrable points, one only has to multiply the δ-function and its transform H by the physical magnitude of the point. In optics there are

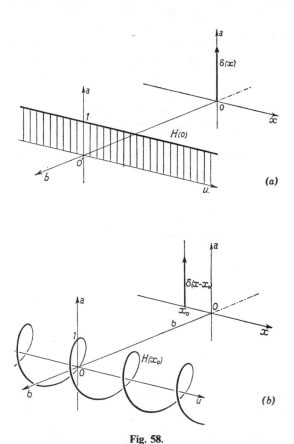

Fig. 58.

also luminous points characterized by their energy $[W\delta(\cdot)]$ or by their complex amplitude $[A\delta(\cdot)]$. Sometimes this is useful and sometimes it is a nuisance. Consider the coefficients of the terms of a Fourier series. They are integrable points, and will be considered later as point functions. Following tradition, we shall continue to omit their symbolic δ, at least when there is no doubt about their dimensions.

A useful representation may be adopted for integrable points. For functions of one variable, the integrated value is represented by a vector, and the infinite density is represented by the line that supports the vector, as illustrated in Fig. 58a and b. The vector alone is often sufficient.

Integrable Points and the Fourier Transform

These two often meet, and we believe that it is the Fourier transform that has introduced the δ-function to "pure mathematics."

Series. The relation that we have established in Chapter 1 between the Fourier integral and series shows that the coefficients of the terms of the series are in fact integrable points. To see this, take the mathematical expression not in its primitive form of Eq. (1.49), but in the definitive notation.

The plane and linear distribution $F(x)$ has the Fourier transform $f(u)$. In optical images, the latter function is always limited. Let U be its range. The terms of the series in x have the period

$$X = 1/U.$$

We call them Z_n, where n is an integer. The equations of correspondence between the integral $F(x)$ and the terms of the series are

$$Z_n = XF(nX), \tag{29}$$

where n is an integer. Z_n has the physical dimensions of the element of integration $F(x)\,dx$. If the δ-function is introduced, Z_n is written, using minimum notation,

$$Z_n = Z\delta(x - nX). \tag{30}$$

It is therefore the integer index n that is equivalent, in the old notation, to the δ-function. That is not surprising. On a graphical scale of numbers, the integers n are the functions $n\delta(x - nX)$. We shall naturally preserve the simpler old notation, which is no longer considered esoteric at any level of mathematical studies.

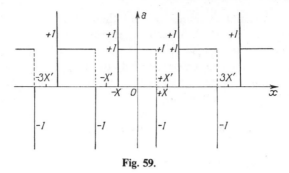

Fig. 59.

Discontinuities. Only one type of discontinuity will be considered. The rectangle function studied in Chapter 3 and illustrated in Fig. 29 has a zero derivative except at the points $-X$ and $+X$. At these two points, the derivative is infinite and has integrable values $+A$ and $-A$:

$$\frac{dF}{dx} = \begin{cases} +A\delta(x + X) & \text{at the point} \quad x = -X, \\ -A\delta(x - X) & \text{at the point} \quad x = +X. \end{cases} \tag{31}$$

Similar forms are found for the derivative dF/dx of the grating function illustrated in Fig. 31 of Chapter 3. A is equal to unity in Fig. 31. Figure 59 shows the derivative dF/dx on the same scale, with out conventions. In both cases, the integration of the derivative yields exactly the function $F(x)$ from which it originated. This is the type of integral that arises in Lebesgue integration, which we mentioned in Chapter 1. Figure 60 illustrates the function $F(x)$ given as an example, which represents the total mass of grains of lead thrown on a plate. The integrable points of the function are of course all real and positive.

General Comment

This just about ends the geometrical properties of the Fourier transform. The geometry is seen to be Euclidian, in the sense that volumes are limited

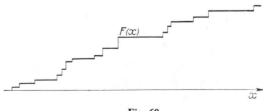

Fig. 60.

by surfaces without thickness, by lines without width, and by points without dimensions. It is strange that those geometrical properties turn out to be compatible with wave optics, but, as we shall see from Chapter 6 on, the geometry of wave distributions is limited by the resolving power, which eliminates the geometrical point. Nonlinear optics may reach the atom but may go no further. This impossibility of molding optics onto a Euclidian geometry has some strange consequences.

5

Applications of Convolution.
Dirichlet's Theorem

PINHOLE CAMERA

Everyone knows how images may be obtained in a dark box with a pinhole camera. For simplification, let the object be in a plane parallel to the image plane and to the plane containing the pinhole, as shown in Fig. 61. The role of diffraction, which will be studied later, will be neglected here.

The correspondence between a point in the object and a point in the image is established by a straight line passing through an arbitrary point in the pinhole aperture. For a point object, the corresponding spot in the image plane satisfies, under most practical conditions, the conditions of isoplanatism in the range of cartesian coordinates where radian measures, sines, and tangents of angles may be considered equal. An appropriate choice of coordinates could make that field still wider.

If the pinhole is small enough, the distribution in the image plane is an image in both the mathematical and the ordinary meaning of the word, in the sense that both isoplanatism and resemblance are present. Suppose the pinhole is larger. The details disappear, but the distribution in the image plane does not cease to depend on all the details of the object distribution, and is still isoplanatic or space-dependent. The resemblance between the object and the image has been lost, but there is still an image in the sense of Parseval's theorem. It is even possible to say that because diffraction effects have been reduced, the image plane receives more information than when there was resemblance.

Let the pinhole plane be replaced by an opaque screen which casts a shadow upon the image plane. The distributions obtained are complementary to the previous distributions, and under the same conditions, satisfy the condition of isoplanatism. There is again an image in the sense of Parseval's theorem. It is impossible to distinguish at first sight the shape of the object in the shape of the shadow, but it also may be calculated.

The example of the shadows illustrates the commutativity of the convolution. Indeed, consider the shadow cast by any object, say a hand, in the light

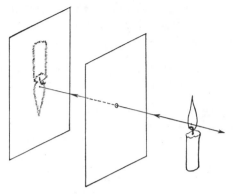

Fig. 61.

of a luminous source, say a lamp. This shadow may be considered as the superposition of the shadows cast by the whole hand for each of the points of the source, and these superpositions create the well-known effects of the penumbra. But then again it may be considered as the superposition of the shadows thrown by each point of the hand for the whole source: if the fingers are moved slightly to leave between them a slight aperture, a rough image of the lamp may be distinguished in the general shadow. In the first case, the spread function corresponds to the hand, and in the second it corresponds to the source.

THE SCANNING SLIT

In order to determine the distribution of energy in a one-dimensional optical image $F(x)$, the image is scanned with a slit of width $2X$, perpendicular to the x-dimension as shown in Fig. 62. The energy integrated by the slit is measured or recorded.

The position of the slit is determined by the position x' of its center. A complete scan of the distribution $F(x)$ yields the distribution

$$D(x) = \int_{x-X}^{x+X} F(x') \, dx'. \tag{1}$$

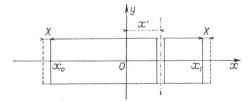

Fig. 62.

The transition between F and D may be considered as an isoplanatic spread where the object is $F(x)$, the effective image is $D(x)$, and the spread function is

$$G(x) = \begin{cases} 1 & \text{for} \quad -X < x < +X, \\ 0 & \text{for} \quad x < -X, \quad +X < x. \end{cases} \qquad (2)$$

The integration must be extended at least over the minimum range containing the object and its image. If p is the length of the object, this range is equal to

$$p' = p + 2X. \qquad (3)$$

The three functions F, D, and G may be represented either by integrals or by series. The expansion of the function $G(x)$ has been seen in the previous chapter. Its spectrum is given by the even function of Fig. 63:

$$g(u) = K \frac{\sin 2\pi uX}{2\pi uX}; \qquad (4)$$

here K is a scaling coefficient, which depends upon the apparatus used.

This is a plane curve; for a given frequency, the terms of F and D therefore have the same phase. There is no phase distortion, but only amplitude distortion. The spectrum $g(u)$ has zeros. The transformation is selective and certain frequencies are removed.

With only one slit, it is impossible to reconstruct the function $F(x)$ completely, the terms with frequencies that are a multiple of $u = 1/2X$

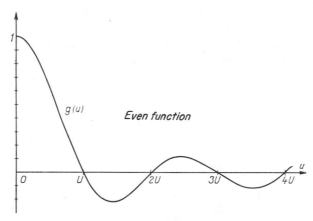

Fig. 63.

having disappeared from the recording, and the neighboring frequencies being too weak for a valid reconstruction. In order to reconstruct the distribution $F(x)$, it is therefore theoretically necessary to have more than one recording made with different slits of properly chosen widths.

Usually a slit much narrower than the range p of the distribution $F(x)$ is chosen. The frequency U for which the first zero of the transfer function appears is then rather high—high enough so that all the terms corresponding to $D(x)$ and $F(x)$ are negligible, lost in the noise. The ideal is for the useful frequencies to be much smaller than U. Because the tangent at the origin of the curve $(\sin t)/t$ is horizontal, and the decrease of the ordinates y is of second order near the origin, the useful amplitudes at low frequencies are approximately equal to unity, and the image is for practical purposes identical to the object. However, in many cases it is useful to take slits of given known selectivities.

The integral (1) may be written, using the definition (2) of $G(x)$,

$$D(x) = F(x) \otimes G(x). \tag{5}$$

In this form it is simple to generalize the definition of the slit and to treat the problem of the slit with a variable transparency or the scanning of a photograph with a distribution of light, and to study the influence of a surface with a variable sensitivity over its area. It is sufficient, as is seen from the mode of integration, that either F or G depend on only one variable. The scanning spot of microphotometers is generally the image of a luminous slit. We shall see that the part of the transmitted light that is not noise and stray light, and that constitutes the ideal scanning spot, has a limited transform with a well-defined frequency. The frequency problem cannot be avoided: the scanning is done frequency by frequency, and not point by point.

RESONANCE CURVE AND IMPULSE RESPONSE

In this section, the conjugate variables will be the time t and the frequency ν. Consider any physical phenomenon described by a nonhomogeneous second-order differential equation. Let the excitation term $F(t)$ be a periodic function with unit amplitude:

$$F(t) = \cos 2\pi\nu t. \tag{6}$$

In the steady-state regime, the system response is another periodic, uniform, unbounded function with the same frequency, but with phase and amplitude distortion:

$$D(t) = Z(\nu)\cos 2\pi\nu t. \tag{7}$$

The factor $Z(\nu)$ is an impedance. In the complex νab space, along the scale of frequency from zero to the infinity that the physicist never attains, may be drawn the resonance curve $Z(\nu)$, which determines the constant coefficients of the differential equation. The resonance curve is shown in perspective in Fig. 64, and its projections on the real and imaginary planes are also shown. For the fundamental frequency ν_0 of the physical system, the modulus of the curve is an absolute maximum. From zero to infinity, the phase goes from zero to π with the fastest change at ν_0.

Any sum of particular solutions is again a solution. Let $f(\nu)$ be an integrable function of ν. The integral

$$F(t) = \int_0^\infty f(\nu)\cos 2\pi\nu t\, dt \tag{8}$$

being the right-hand member of the differential equation, the response is obviously

$$D(t) = \int_0^\infty f(\nu)z(\nu)\cos 2\pi\nu t\, d\nu. \tag{9}$$

$D(t)$ is an image of $F(t)$. The cosine transform of $F(t)$ is $f(u)$, and the cosine transform $d(\nu)$ of $D(t)$ is, according to Eq. (9),

$$d(\nu) = f(\nu)z(\nu). \tag{10}$$

The function $z(\nu)$ is therefore the cosine transform of the spread function

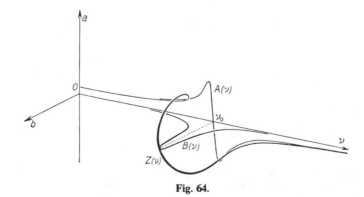

Fig. 64.

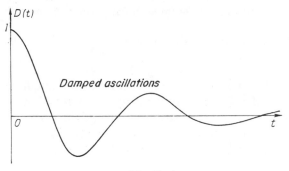

Fig. 65.

$Z(t)$, which is the operator that transforms the object $F(t)$ into the image $D(t)$.

What has been said previously regarding convolutions involving the delta function may be applied to the calculation of $Z(t)$. Let the input be a delta function at $t = 0$:

$$f(\nu) = 1, \qquad F(t) = \delta(t). \tag{11}$$

If T is the Fourier-transform operator, then

$$T[D(t)] = z(\nu), \tag{12}$$

$$D(t) = Z(t). \tag{13}$$

Now $Z(t)$ is the response to the impulse $\delta(t)$. The excitation is an amplitude impulse, which means that the response begins at time zero with unit amplitude. For a galvanometer, this would be a free oscillation starting with a unit deviation. The response is represented in Fig. 65: it is a damped oscillation, the exact Fourier transform of the resonance curve of Fig. 64.

What we call a transmission function in optics is an impedance here.

DIRICHLET'S THEOREM—WEIGHTING FUNCTIONS

Dirichlet's theorem may be stated in the following general way:

Let a function $F(x)$ defined at each point over its range have as its only singularities a finite number of discontinuities of the first and second kind, that is, representable by a Fourier series or integral. The value of $F(x)$ at a *discontinuity is the average of the immediately preceding and following values*:

$$F(x) = \tfrac{1}{2}[F(x + \varepsilon) + F(x - \varepsilon)] \qquad \text{for} \quad \varepsilon \to 0. \qquad (14)$$

The theorem yields the value of the function at the amplitude and phase discontinuities. Take for example the rectangle function of Fig. 29 of Chapter 3; its definition disregards the values at the two discontinuities at $-X$ and $+X$. Indeed, those values play no role in the transformation integrals. They do not even appear in the definition of the derivative

$$\frac{dF(x)}{dx} = A[\delta(x + X) - \delta(x - X)]. \qquad (15)$$

But they are determined by Dirichlet's equation:

$$F(-X) = F(X) = \frac{A}{2}. \qquad (16)$$

In many figures of Chapter 3, the values at the discontinuities are represented by a big circular dot. These values play no role in the calculations of the transforms, because they have an identically zero differential element and correspond to a finite value of the function. However, they may be calculated, sometimes with considerable effort, by summing the Fourier integral or series at the discontinuity.

This may be extended to functions of more than one variable. In Fig. 66 is shown the two-dimensional equivalent of the rectangle function; the nonzero area is cross-hatched. It is a magic carpet, anchored by the rectangle of Dirichlet's points.

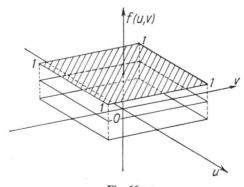

Fig. 66.

Extension of the Theorem

Equation (14) is the starting point of a convolution where $F(x)$ is dissipated by the spread function

$$G(x) = \frac{\delta(x + \varepsilon) + \delta(x - \varepsilon)}{2} \tag{17}$$

The Fourier transform of $G(x)$ is

$$g(u) = \frac{H(+\varepsilon) + H(-\varepsilon)}{2} = \cos 2\pi\varepsilon u. \tag{18}$$

The Fourier transform of the convolution

$$F'(x) = F(x) \otimes G(x) \tag{19}$$

is

$$f'(u) = f(u) \cos 2\pi\varepsilon u. \tag{20}$$

When ε tends to zero,

$$F'(x) \equiv F(x),$$
$$\tag{21}$$
$$f'(u) \equiv f(u).$$

The theorem is thus proven simultaneously for all the points of the function $F(x)$. All this is classical.

Weighting Functions

It is of interest to substitute some continuous functions in Eq. (19), at the two integrable points of Eq. (17). We shall try four functions.

The Rectangle Function. The spread function is equal to

$$G(x) = \begin{cases} \dfrac{1}{2\varepsilon} & \text{for} \quad -\varepsilon < x < +\varepsilon, \\ 0 & \text{elsewhere.} \end{cases} \tag{22}$$

The function $G(x)$ is normalized to unity. To calculate $F'(x)$ we use the second procedure for image formation, remembering that $G(x)$ is a real and even function:

$$F'(x) = \frac{1}{2\varepsilon} \int_{x-\varepsilon}^{x+\varepsilon} F(x)\, dx$$

$$= \overline{F(x)} \text{ between } x - \varepsilon \text{ and } x + \varepsilon. \tag{23}$$

In the limit, when ε tends to zero, $G(x)$ tends to $\delta(x)$ and its transform $H(\theta)$ tends to unity. Once again the identities of Eq. (21) are found. It is therefore possible to state that

The value of the function $F(x)$ at x is equal to the limit of its mean value in the range $x - \varepsilon$ to $x + \varepsilon$ when ε tends to zero.

The Gaussian Function.   Let $G(x)$ be the normalized function

$$G(x) = \frac{a}{\sqrt{\pi}} \exp(-a^2 x^2), \tag{24}$$

$$g(u) = \exp\left(-\frac{\pi^2}{a^2} u^2\right). \tag{25}$$

Then

$$\int_{-\infty}^{+\infty} G(x)\, dx = 1 \quad \text{and} \quad g(0) = 1. \tag{26}$$

As the parameter a goes to infinity, $G(x)$ of course again tends to the function $\delta(x)$, and $g(u)$ tends to $H(0) = 1$, which again brings about the identities of Eq. (21). However, in this case the convolution equation

$$F'(x) = \int_{-\infty}^{+\infty} F(x') \frac{a}{\sqrt{\pi}} \exp\left(\left[-a^2(x - x')\right] dx'\right), \tag{27}$$

which is necessarily extended to the whole range of $F(x)$, makes it more obvious that all the points of F enter into the definition of F'. The importance of this extension is shown in the two examples that follow.

Eppler's Faltung and the Time Techniques.   The *faltung* is here defined as the convolution integral stopped at the instant t:

$$J(t) = \int_{-\infty}^{+\infty} F(t') G(t - t')\, dt'. \tag{28}$$

Let us speak in the language of time. The variable t' is the past of the instant t. It is an accumulation of memories. On the other hand, t remains the time variable—the instant, a cut, a discontinuity. It is, when all is said, the present instant. Faltungs thus understood are the usual forms of recordings of phenomena. $F(t)$ is the real phenomenon, the flash photograph; $J(t)$ is the recorded phenomenon, a rather special image for which $G(t)$ characterizes the method and the recording apparatus, and carries the systematic memory of the recent past, by means of an extemporaneous mechanism.

Let the instant t be the origin of time, positive values going towards the future:

$$J(0) = \int_{-\infty}^{0} F(t')G(-t')\, dt', \qquad \text{with} \quad t' \leq 0. \qquad (29)$$

By definition, t' is always negative, and $-t'$ is always positive. Let

$$\theta = -t'. \qquad (30)$$

where θ is always positive and is thus oriented exclusively towards the future, the future seen by F at the past instant t', as illustrated in Fig. 67. Each ordinate M of the past $F(t')$ is the origin of a particular function of influence $F(t')G(\theta)$ which carries, at the present instant $t = 0$, the attenuated influence of the past. This is just the first mode of image formation seen earlier.

Let the second mode of image formation now be considered. The product $F(t')G(-\theta)$, integrated over the whole past, yields the whole influence of the past on $F(0)$, that is to say $J(0)$. This is illustrated in Fig. 68.

A priori, the function $F(t)$ is integrable only over an arbitrary limited range. This is the case for meteorological recordings or for the physics of the earth, and for many electrical and electronic measurements. It is thus necessary that convergence be assured by a convergent weighting function. In Figs. 67 and 68, $G(\theta)$ has the shape of a damped exponential. In some cases G is a constant, usually equal to unity, over some limited time range T. In meteorology, this is generally taken to be the average over the past ten years.

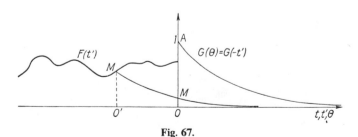

Fig. 67.

The variable t of Eq. (28) and the zero of Eq. (29) and of Figs. (67) and (68) are not discontinuities for $F(t')$, $J(t)$, $J(0)$, and $G(\theta)$. They are ends in the full sense of the word, being both goals and conclusions for the observer. It is impossible to define them at values both immediately before and immediately after: for F and J, the value after is missing; for G, the value before. Only the extemporaneous weighting function $G(t)$ shows the evolution of the theorem: the two original ε approaching zero is replaced by the condensation of the function $G(\theta)$ about its origin into a δ-function. The function J then becomes identical to F.

The values of $J(\cdot)$, $F(\cdot)$ and $G(\cdot)$ in Eq. (29) can never be attained, and the values of $J(0)$, $F(0)$, and $G(0)$ only exist in the limit.

In techniques of electrical recording, the damped exponential function $e^{-\theta/\tau}$ appears repeatedly. Smaller values of the time constant τ has been striven for since the beginning of the century. At best, it was then of the order of a millisecond. Today, electronics operates in the realm of nano-seconds.

Bandlimited Functions. In the following chapters, it will be shown at length that the Fourier transforms of distributions of light in one, two, or three dimensions, with complex amplitudes or distributions of energy, remain confined in the range of low spatial frequencies smaller than $2/\lambda$. Everything that involves band-limited functions is therefore important for optics. However, we shall again restrict ourselves to mathematical proper-ties.

Functions of One Variable

Let the Fourier transform $f(u)$ of a function $F(x)$ have a finite range centered on the origin $u = 0$:

$$f(u) = 0 \quad \text{for} \quad u < -U, \quad +U < u. \tag{31}$$

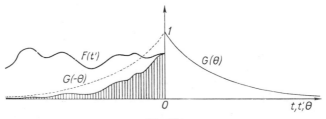

Fig. 68.

Let $G(x)$ be a weighting function whose Fourier transform is also zero outside a certain range:

$$g(u) = \begin{cases} 1 & \text{for} \quad -U' < u < +U', \\ 0 & \text{elsewhere.} \end{cases} \tag{32}$$

The function $G(x)$ is normalized and equal to

$$G(x) = 2U' \frac{\sin 2\pi U' x}{2\pi U' x}. \tag{33}$$

Normally, if U' were increased indefinitely, $g(u)$ would tend to the function $H(0)$ and therefore $G(x)$ would tend to $\delta(x)$; but it is not possible to go so far as that. If

$$F'(x) = F(x) \otimes G(x), \tag{34}$$

then

$$f'(u) = f(u)g(u), \tag{35}$$

and the range of $f'(u)$ is restricted to the smaller of the ranges of f and g. Because the range of $f'(u)$ is unalterable, there is no point in making $U' < U$. For

$$U' \geqslant U \tag{36}$$

we have

$$f(u)g(u) \equiv f(u) \tag{37}$$

and

$$F(x) \otimes G(x) \equiv F(x). \tag{38}$$

This is equal to

$$F(x) \otimes 2U' \frac{\sin 2\pi U' x}{2\pi U' x} = F(x). \tag{39}$$

Let φ be the smallest extent of the range of $f(u)$, and $\varphi(u)$ the value of $g(u)$ corresponding to the range $U' = U$. The preceding expressions become

$$f(u)\varphi(u) = f(u), \tag{40}$$

$$F(x) \otimes \Phi(x) = F(x), \tag{41}$$

$$F(x) \otimes 2U \frac{\sin 2\pi U x}{2\pi U x} = F(x). \tag{42}$$

Now because $F(x)$ is a continuous function, the initial classical form of Dirichlet's theorem given by Eq. (14) is obviously still applicable, but when it is expressed in the form of Eq. (42), it is much more significant. $F(x)$ is not distributed point by point, but by extended and even unbounded functions $\Phi(x)$. The point differential element $dF(x)$ is in reality the convolution element

$$d\left[F(x') \otimes \Phi(x' - x)\right]. \tag{43}$$

The preceding proof, which assumed that all the ranges of the functions had their centers at the origin, is still valid for arbitrary limits of φ. The function may be reduced to the preceding case without changing the distributions of amplitudes by a translation of $\varphi(u)$, accompanied by a helical twist (that is, a linear phase shift) of $F(x)$.

This theorem is so characteristic of optical imagery that we now give a very general proof, valid for functions of more than one variable.

Let F, an integrable function that is zero outside a finite range φ, have a Fourier transform f. Let a second function G of the same variables as F with Fourier transform g be equal to unity at each point over a finite but variable range ψ, which is initially contained in the range

$$\psi \in \varphi. \tag{44}$$

To the algebraic product of the two transforms,

$$fg = f', \tag{45}$$

there corresponds the convolution

$$F \otimes G = F', \tag{46}$$

where the range of F' is the common part φ' of the two ranges φ and ψ.

The range ψ may be extended until it coincides exactly with φ. Then G and g take the values G_0 and g_0, and Eqs. (45) and (46) become

$$fg_0 = f, \tag{47}$$

$$F \otimes G_0 = F. \tag{48}$$

The convolution now yields a function that is identical to the function F.

If the range ψ is now extended beyond the limits of the range φ, Eq. (47) is still valid, where g_0 represents the common part of the two ranges. G must be substituted for G_0, but the convolution product $F \otimes G$ again yields F.

In order to simplify the notation, it is useful to represent the finite ranges by "plateau" functions with value unity over the ranges φ under consideration. Thus we define the function

$$\varphi(\cdot) = \begin{cases} 1 & \text{over the range of } f, \\ 0 & \text{elsewhere.} \end{cases} \tag{49}$$

In the space of F, the Fourier transform of this function is $\Phi(\cdot)$. The final form given to our two products is therefore

$$f\varphi(\cdot) \equiv f, \tag{50}$$

$$F \otimes \Phi(\cdot) \equiv F. \tag{51}$$

In summary, any band-limited function F—that is, any function whose Fourier transform is contained in a finite range φ—may be considered as resulting from the convolution of itself with the plateau function that is unity inside the limits of φ, and zero outside.

There is no major inconvenience in using the same letter to designate the range and its plateau function. We call the function $\varphi(\cdot)$ the *range function*, and the spread function $\Phi(\cdot)$ (internal diffraction function) of the function F we call the *internal correlation function*.

It will be seen at length in the following chapters that this peculiarity of band-limited functions is an essential property of light distributions, and is justified by wave theory. More generally, it is justified by linear optics.

Multiple Convolutions

Consider again Eqs. (15) and (16) of the preceding chapter:

$$F_0 \otimes G_1 \otimes G_2 \otimes \cdots \otimes G_{n-1} = F_n, \tag{4.15}$$

$$f_0 g_1 g_2 \cdots g_{n-1} = f_n. \tag{4.16}$$

Let the functions in Eq. (4.15) be called *components*, and the functions in Eq. (4.16) be called *spectra*. A third expression involving the ranges of the supports of the functions may be added to those two:

$$\varphi_0 \varphi_1 \varphi_2 \cdots \varphi_{n-1} = \varphi_n, \tag{52}$$

The corresponding relation for the internal correlation functions is

$$\Phi_0 \otimes \Phi_1 \otimes \Phi_2 \otimes \cdots \otimes \Phi_{n-1} = \Phi_n. \tag{53}$$

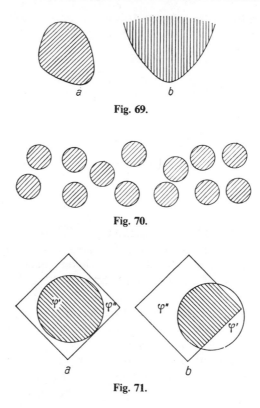

Fig. 69.

Fig. 70.

Fig. 71.

In optical images, and more generally in linear optics, the variables of the components are the three cartesian coordinates of space, x, y, and z, plus (occasionally) the time t. The conjugate variables are therefore the spatial frequency variables u, v, w, plus the conjugate variable s of time. The ranges are thus bounded by points, lines, and surfaces, and if time is involved, by volumes. Let us leave the latter exceptional case aside. We have drawn up the following corollaries for the case of plane distributions of monochromatic light. They are an immediate illustration of the Fourier transform and convolution. They may easily be generalized.

1. The zeros of the spectra and of the range functions are all present in the spectra and in the range functions of the final component f_n.
2. If all the spectral ranges of the components are unlimited, then the spectral range of the resultant is unlimited.
3. If only one of the spectral ranges φ' is limited, the range φ_n has the same limit:

$$\varphi_n = \varphi'. \tag{54}$$

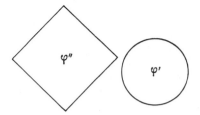

Fig. 72.

In particular, if only one of the transmissions by convolution is done on a linear optical image with frequency $\mu = 1/\lambda$, the final range of frequencies will be smaller than $\mu/2$.

4. The limit of the range may be a closed curve as in Fig. 69a, or an open curve as in Fig. 69b; the plateau functions are hatched in the figures.

5. An unlimited or limited range may be disconnected, as shown in Fig. 70.

6. If two spectral ranges φ' and φ'' are limited, they only transmit their common range; this is illustrated in Fig. 71a and b.

7. If two limited ranges have no common part, then the area of Φ_n is equal to zero, and f_n and F_n are identically equal to zero. This is illustrated in Fig. 72.

6

Fraunhofer Diffraction

INTRODUCTION

We have now established the major part of our required mathematical apparatus: the diffracting function and its spectrum are a Fourier-transform pair; to the convolution relationship between the object and its image corresponds a product between the spectra in the spectral domain. It only remains to develop the conventions for expressing this. After that, the path opened to us by the definition of the image will lead us towards optical transmission.

Before following this path, the author would like to give a perspective view of a countryside where he once browsed as a sightseer.

A reversible or partially reversible transmission has, as in thermodynamics, two different forms:

1. The object and the image may be considered as an initial and a final state, or if one prefers, an initial and a final transformation, like the heat source and the heat sink of Carnot. The mode of transmission remains unknown or undetermined, like Carnot's "transporting fluid." It is only required that the transport be adiabatic and at least partially reversible. This represents the point of view that all theories of images are also theories of objects. Convolution theory yields a model, and incoherent imagery yields a domain. This type of transmission may be said to be controlled from the exterior, by means of transformation invariants between the object and image spaces.

2. In the second form, a mode of transmission is defined that may be progressively integrated from the object to the image, and conversely, for the reversible part of the image, from the image to the object. The transmission is said to be controlled from the interior. In thermodynamics, the nature and the properties of the transporting fluid are defined by equilibrium diagrams, by equations of state. This is the role played in optics by equations of propagation: ray-tracing equations, Fermat's principle,

Huyghens's principle, wave equations, and the laws of interaction between matter and light.

The Fourier transform was first introduced into incoherent imagery. Convolution and the thermodynamic transformation of the first type were perfectly suited for this. On one side there is an object, or more exactly, a representation of an object, and at the other end of the simple transmission link, an image which is required to resemble the object as much as possible. The object is a material object, either illuminated or self-luminous. It may be defined point by point according to euclidian geometry, or at least atom by atom, which is equivalent in this case.

However, the theory of the image of a point object must be consistent with the wave theory. Calculated by means of the complex amplitudes, the distribution of energy over the image is determined by classical means. This chapter begins with this calculation. The plane, integrable complex amplitude $F(x, y)$ has a Fourier transform which is its Fraunhofer diffraction pattern, or its diffraction pattern at infinity $f(u, v)$, where u and v are direction cosines. Thus an unlimited distribution in the xy plane is related to an angular distribution among the directions of propagation through the plane. This is the progressive propagation law that allows the study of the second mode of transmission mentioned above.

When Abbe's condition is derived from the Fourier transform and convolution, this law of propagation is related to the theory of isoplanatic images, in a manner very similar to the theory of incoherent images of the previous chapter, but extended here to complex amplitudes of images illuminated with coherent light. However, the old habit of defining a priori the integrable distribution $F(x, y)$ over diaphragms and obstacles that limit the beam, will have to be abandoned in favor of defining the distribution in free propagation by its Fourier transform $f(u, v)$ inside the limited range of the circle of extension.

The author has related this circle to the Laue-Epler sphere, which plays a similar role in x-ray diffraction. As an excuse for this intrusion, he looks to the linear optics of photon-matter interactions, which are found in nonlinear optics. This is related to the structure of matter and its optical properties, and to the passage of photons through this structure, questions which are of the realm of the three-dimensional world.

Linear optics happily accepts nonlinear phenomena among its fundamental phenomena: reflection, refraction, transmission through transparent media or through thin films of metal. Their nonlinearity appears in the polarization of light, in molecular or atomic diffraction, and in absorption. The instrumental optician is obsessed with the homogeneity of media:

highly polished surfaces, highly exact geometrical surfaces, to the accuracy of one atom if possible. Phenomena of nonlinear optics are accepted when they yield linear beams of photons.

Diffraction of a One-Dimensional Plane Wave

The relation between diffraction phenomena at infinity and the Fourier transform was pointed out in its simplest form by Michelson in 1891, in one of his famous articles on the analysis of fine spectral lines. Michelson gave the equations, and in a letter published the following month, Lord Rayleigh mentioned Fourier's name. This is the origin of the theorems whose consequences will be developed. This initial form will be proven first.

Consider an infinite train of plane waves parallel to the plane xOy and propagating in the direction of the normal ON as illustrated in Fig. 73. Let an absorbing screen with a density depending only on x be placed in the plane xOy. The transmission factor is a constant along any line parallel to Oy. The extension of the screen along the y-axis is assumed to be great enough so that diffraction along Oy is negligible. The amplitude of the wave leaving xOy is given by the real function $G(x)$. The resultant of the amplitude $g(i)$ propagating to infinity in the direction OM from the plane xOy may be calculated by applying Huyghens's principle to the wave amplitude at the plane xOy. If ω is the angular frequency, λ is the wavelength, and i is the angle NOM (whose sine is positive when x is positive), then the diffracted amplitude is

$$g(i) = \int_{-\infty}^{+\infty} dx \int_Y G(x)\cos\left(\omega t - \frac{x \sin i}{\lambda}\right) dy. \qquad (1)$$

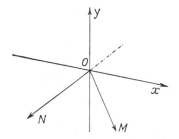

Fig. 73.

If the first integration of $G(x)$ with respect to y is assumed to yield a real function $F(x)$, λ is normalized to unity, and

$$\sin i = u, \tag{2}$$

then Eq. (1) may be written

$$g(u) = \int_{-\infty}^{+\infty} F(x)\cos(\omega t - 2\pi ux)\, dx. \tag{3}$$

$$G(u) = \cos \omega t \int_{-\infty}^{+\infty} F(x)\cos 2\pi ux\, dx + \sin \omega t \int_{-\infty}^{+\infty} F(x)\sin 2\pi ux\, dx. \tag{4}$$

When complex exponentials are used to simplify the trigonometrical calculations, Eq. (4) may be written

$$\exp(i\omega t)\, f(u) = \exp(i\omega t)\int_{-\infty}^{+\infty} F(x)\exp(-i2\pi ux)\, dx. \tag{5}$$

The periodic time dependence $\exp(i\omega t)$ cancels out, and $f(u)$ is the Fourier transform of $F(x)$:

$$f(u) = {}^*T[F(x)]. \tag{6}$$

The reciprocal relation is therefore

$$F(x) = T[f(u)]. \tag{7}$$

Note that because $F(x)$ is real, the transform is symmetrical with respect to the real axis. Michelson proposed using this method to obtain the visibility curve of the fringes given by his interferometer in the presence of a complex spectral line. Our x is the variable wavelength in the profile of this complex spectral line, and u is the path difference between the two interfering waves.

Pupils with Two Variables

The complex notation will now be used exclusively. The wave motion may be described by a modulus $\rho(x, y)$, a phase $\varphi(x, y)$, and a periodic time variation $\exp(i\omega t)$, where ω is the angular frequency:

$$\rho(x, y)\exp i[\omega t + \varphi(x, y)] = \exp(i\omega t)\, F(x, y), \tag{8}$$

where

$$F(x, y) = \varphi(x, y)\exp[i\varphi(x, y)] \qquad (9)$$

is called the complex amplitude of the wave.

Let Oz be the normal to the plane xOy, and OM be a direction of propagation defined by the projections u and v of a unit vector pointing in a direction opposite to the direction of propagation of the light. This is shown in Fig. 74. The state of the wave at infinity in the direction OM may be deduced from the state of the wave in xOy by means of Huyghens's principle, where each element of the plane xOy has a complex amplitude $F(x, y)$.

The wave leaving a point P with coordinates x, y, O and propagating along PM' parallel to OM has a phase lag of $2\pi(ux + vy)/\lambda$ with respect to a wave leaving point O. This phase lag is normalized to $2\pi(ux + vy)$. In the direction OM, the total amplitude at infinity is given by the integral

$$\exp(i\omega t)\int\int_{-\infty}^{+\infty} F(x, y)\exp[-i2\pi(ux + vy)]\, dx\, dy. \qquad (10)$$

The factor $\exp(i\omega t)$ represents the wave motion; from now on it will be dropped. In the plane at infinity, the complex amplitude may be expressed in terms of the angular coordinates u and v as the Fourier transform

$$f(u, v) = \int\int_{-\infty}^{+\infty} F(x, y)\exp[-i2\pi(ux + vy)]\, dx\, dy. \qquad (11)$$

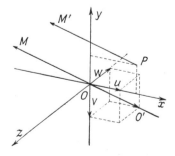

Fig. 74.

This transformation is of the type (2.33), which has the reciprocal relation similar to (2.34)

$$F(x, y) = \int\int_{-\infty}^{+\infty} f(u, v)\exp[i2\pi(ux + vy)] \, du \, dv, \qquad (12)$$

In abridged form, the two relations are written

$$f(u, v) = T^*[F(x, y)],$$
$$F(x, y) = T[f(u, v)]. \qquad (13)$$

Coming back to one-dimensional transforms, Eq. (11) may be written

$$F(x, y) = \int_{-\infty}^{+\infty} \exp(i2\pi ux) \, du \int_{-\infty}^{+\infty} f(u, v)\exp(i2\pi vy) \, dv. \qquad (14)$$

Along the lines v = constant, the transformation is reduced to a one-dimensional transform of the variable u.

In sum, *the diffraction pattern of a plane pupil at infinity is the Fourier transform of the complex amplitude distribution on the pupil plane.*

The reciprocal nature of the equations (13) corresponds to what is called *inverse diffraction*.

Reciprocity and Inverse Diffraction

Consider Eq. (11) and Eq. (12), which are summarized in Eq. (13). Because x and y are linear variables on a plane, it is natural to give them an unlimited extension. It is true that real pupils are limited, and it has become habitual to see this as the cause of diffraction. In elementary treatises on instrumental optics, diffraction is always associated with the finite extension of beams of light. From a mathematical point of view, there is nothing awkward here: the function $F(x, y)$ must be defined over the whole plane; if the function is limited, it is simply defined as being equal to zero outside its range.

On the other hand the variables u and v are direction cosines, which by definition are restricted to the frequencies inside the circle with a radius equal to unity:

$$u^2 + v^2 < 1. \qquad (15)$$

The function $f(u, v)$ may not be periodic and is restricted to its circular range. This raises a difficulty in the parallel development of Fourier theory and of diffraction phenomena. This difficulty is usually avoided by considering only cases where $f(u, v)$ becomes negligible before reaching the edge of its circle of extension. In such cases a negligible fraction of the diffracted energy is diffracted at large angles. The limits of integration over the variables u and v are then fictitiously extended to infinity.

This approximate physics is unacceptable in the case of gratings. In fact the extension of the variables u and v requires a three-dimensional geometry and a third direction cosine w. It is not an independent variable; the relation between direction cosines

$$u^2 + v^2 + w^2 = 1 \tag{16}$$

suggests that $f(u, v)$ should be defined on a sphere. Such a solution will be considered at the end of this chapter.

If we succeed in making the Fourier transform correspond to the physical phenomenon while satisfying the conservation of energy, the principle of inverse diffraction will be proven. The direction of propagation appears nowhere in the Fourier transform, and in Fourier theory, inverse diffraction is called *reciprocity*.

Infinity

When one of his students asked Raymond Boulouch, "Excuse me, Sir, but where is infinity?" he used to answer, "Here, it is at 25 meters." And he pointed to the other end of the long gallery where we did our experiments. The infinity of optics is not the same as the infinity of mathematics. The infinity at 25 meters was perfectly sufficient to adjust the small sights of our Pellin goniometers. We shall make a few remarks about this infinity, the nature of which is further complicated by the development of particle optics.

In coherent imagery, the plane at infinity is generally the focal plane of a convergent lens located after the pupil. The centered system that practitioners of filtering often call the double diffraction system is illustrated in Fig. 75a.

A coherent plane wave is produced by a monochromatic point source S at the focus of a lens O_1'. The plane xOy of the pupil is at P_1. A diaphragm, absorbing screens, or variable-index or variable-thickness plates placed between the lens O_2' and the plane P_1 introduce deformations into the plane wave; the complex amplitude at P_1 is equal to $F(x, y)$, a limited function.

In order to observe $f(u, v)$ at a finite distance, a lens O_2'' is placed with its image focal plane at P_2. The light propagating from P_1 in the direction OM converges to the point M. Only regions of the plane P_2 where the lens

O_2'' has a field that is sufficiently plane and undistorted may be observed correctly. We are therefore limited to the low frequencies of $F(x, y)$.

In order to attain without distortion the higher frequencies of $f(u, v)$, the axis of O_2'' must be inclined relative to the direction SN by a rotation about the point O. This is illustrated in Fig. 75b. The whole distribution of energy on the sphere with center O could be investigated by recording the light passing through a small aperture at the focus M of the lens O_1''. Let this spherical distribution be $f(u, v, w)$. The corresponding energy distribution is $|f(u, v, w)|^2$. The phases could be found by phase contrast or holography.

These methods of focusing have an important property. The distribution $f(u, v)$ present in the plane P_2 is a surface with wave properties just like $F(x, y)$. It is a pupil if P_2 coincides with a second lens O_2 similar or identical to lens O_1, the combination of O_1' and O_1''; this pupil P_2 will yield a diffraction pattern $F'(x, y)$ in the plane P_3, which is a filtered image that may be deformed by what is contained in the plane P_2 of $F(x, y)$. We could continue.

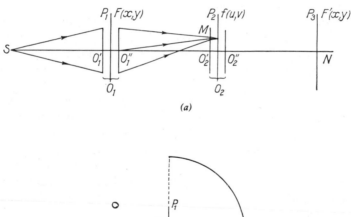

(a)

(b)

Fig. 75.

If we really go to infinity, things become more complicated. It must be remembered that infinity is not a place but a concept: it is always approached but never attained. As the distance between the pupil and the observer increases, two essential phenomena happen:

1. Because the pupil is of necessity limited, the degree of coherence increases and tends to unity; it becomes unresolvable and satisfies the coherence condition of Young and Fresnel, which in our opinion is the fundamental rule of coherence.

2. Because $f(u, v)$ is a fan of divergent directions, the photons eventually become resolvable. The more coherent they are, the more they are separable; it is sufficient that they be separated by distances greater than $\lambda/2$. Their coherence is a result of their trajectories being almost parallel: they seem to come from the same point.

The coherence of those random distributions of separable photons has an origin distinct from that which distinguishes their nearby coherent and incoherent pupils, which are characterized by whether or not the phase of the amplitude distribution has a well-defined relation from point to point. In the immediate vicinity of the material of the diaphragms, diffraction fringes originate which invade the beam and progressively organize themselves into the fringes at infinity. It is not easy to see, through those fringes, the trajectories of the photons that carry the energy, and that propagate in straight lines after being broken up in the xy plane. A simple example of such a system of fringes is shown in Fig. 76. A Fresnel biprism $P'P''$ is illuminated by a parallel beam of light, S, which is split into two beams S' and S'' having in common the volume shown in the figure as the rhombus $OABA'$. Because the apex of the prism is perpendicular to the plane of the figure, the interference fringes between the two beams are plane and perpendicular to the same plane. Sections of those fringes are illustrated. With good optics, a contrast of unity may be obtained. With a 60-mm-wide Senarmont birefringent quartz prism, the author and Jacques Clerc have produced a complete triangle of fringes with a base $P'P''$ whose length was about 20 meters.

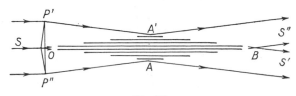

Fig. 76.

All the photons go through the dark fringes where their energy is unavailable, suggesting that Newton's explanation may be inadmissible.

THE PHYSICAL MEANING OF THE FOURIER TRANSFORM

The Refractive Index

Let a plane P separate two media having different indices of refraction. Let a perpendicular plane be the plane of incidence. Along its intersection with the plane P, the complex amplitude in medium 1 has the complex amplitude $F(x)$; the wavelength λ, is taken as the unit of length. In medium 2, the unit of length is λ', and the new abscissa is x'. Let n be the relative refraction index between medium 2 and medium 1:

$$\lambda = n\lambda',$$

$$x' = nx. \tag{15}$$

The new distribution of complex amplitudes on side 2 is found by a simple change of variables in $F(x)$, except for a proportionality constant A', which comes from the conservation of energy:

$$F(x) = A'F'(x') = A'F'(nx). \tag{16}$$

The Fourier transform of $F'(x')$ is equal to

$$f'(u') = \int_{-\infty}^{+\infty} F'(x')\exp(-i2\pi u'x')\, dx'. \tag{17}$$

Carrying out the change of variables of Eq. (15) yields

$$f'(u') = \frac{n}{A'} \int_{-\infty}^{+\infty} F(x)\exp(-i2\pi nu'x)\, dx,$$

where $f(u)$ is the Fourier transform of $F(x)$;

$$f'(u') = \frac{n}{A'} f(nu').$$

The functions $f'(u')$ and $f(u)$ are scaled versions of each other, with the variables

$$u = u', \tag{20}$$

$$\sin i = n \sin i'. \tag{21}$$

   This is Descartes' law of refraction. Of course we could have postulated this law as an experimental fact and derived the change of variables from it, but we wish to insist on the role that the wavelength plays as a unit of measurement.

Abbe's Condition

Abbe's condition is identical to the condition for isoplanatism. An independent proof is given here in order to show the connection between the Fourier transform and Abbe's condition in coherent and incoherent imagery.

   Let $F(x, y)$ be a plane distribution and $F'(x', y')$ be its image. In both the object and image spaces, the unit of length is the wavelength. The relation between indices and wavelengths is a relation between units:

$$n\lambda = n'\lambda'. \tag{22}$$

The object and image distributions both have Fourier transforms:

$$T^*[F(x, y)] = f(u, v), \tag{23}$$

$$T^*[F'(x', y')] = f'(u', v'). \tag{24}$$

   To say that $F'(x, y)$ is an image of $F(x, y)$ is to say that it is derived from $F(x, y)$ by a systematic deformation. The properties of radiating energy suggest describing the deformation by means of a relation between the conjugate variables u, v, u', v':

$$u' = \varphi(u, v) \quad \text{and} \quad v' = \psi(u, v). \tag{25}$$

which defines a transmission factor such that

$$f'[\varphi(u, v), \psi(u, v)] = f(u, v)h(u, v). \tag{26}$$

   The conditions must be determined under which a *translation without deformation* of the distribution $F(x, y)$ with components a and b along Ox and Oy results in a *transformation without deformation* of $F'(x', y')$ with components $O'x'$ and $O'y'$. For the shifted functions, the shift theorem of Fourier theory yields

$$T^*[F_1(x, y)] = f_1(u, v) = f(u, v)\exp[i2\pi(ua + vb)], \tag{27}$$

$$T^*[F'(x', y')] = f'(u', v') = f'(u', v')\exp[i2\pi(u'a' + v'b')], \tag{28}$$

which lead to a new transmission factor

$$h_1(u, v) = h(u, v)\exp i2\pi[(u'a' + v'b') - (ua + vb)]. \qquad (29)$$

For the conjugate planes xy and $x'y'$, a sufficient condition for Eqs. (23) and (24) to be valid is

$$h_1(u, v) \equiv h(u, v), \qquad (30)$$

which requires

$$(u'a' + v'b') - (ua + vb) = N \text{ (integer)}. \qquad (31)$$

The solution $N = 0$ corresponds to the usual single image. The solutions $N \neq 0$ correspond to the multiple images of gratings. For the pairs of conjugate axes $Ox\text{-}O'x'$ and $Oy\text{-}O'y'$, Eq. (31) is equivalent to two relations

$$u'a' - ua = N \quad \text{or} \quad u'x' - ux = N \text{ (integer)};$$
$$\qquad (32)$$
$$v'b' - vb = M \quad \text{or} \quad v'y' - vy = M \text{ (integer)}.$$

For centered systems without anamorphosis, the magnifications for the x and the y directions are equal, and N and M are equal to zero. The relations then collapse into a single relation, Abbe's sine condition

$$ux = u'x', \qquad (33)$$

which in the classical notation is equivalent to

$$nx \sin i = n'x' \sin i'. \qquad (34)$$

The author has often wondered why such ideal images are not called *abbeian*, instead of being afflicted with the barbarous adjective isoplanatic.

The Law of Propagation and the Circle of Extension. It is appropriate at this point to apply these rather abstract concepts to a concrete example. The simplest example, due to Raymond Boulouch, is the planar refraction illustrated in Fig. 77.

The plane D separates two homogeneous media having indices of refraction equal to unity and to $n > 1$ respectively. Consider an object plane parallel to D in the medium of index n. For each object point A, there corresponds, in the image space of D, a beam with a caustic having circular symmetry whose axis is the line AA' perpendicular to D. For point A, this caustic has a linear sagittal surface along AA' and a tangential surface represented by its meridian line.

A translation from A to B shifts all the rays emanating from A or B. From the symmetry of the problem, the image must have circular symmetry, and the center, or rather the axis, of the image must be along AA' for A and along BB' for B. The magnification is equal to unity; Abbe's condition is satisfied and is reduced to Descartes' law:

$$u' = nu \quad \text{and} \quad v' = nv,$$
$$\sin i' = n \sin i'. \tag{35}$$

Consider the beams themselves, independently of any consideration about images. It is sufficient to consider the beam from point A. This beam has circular symmetry about the axis AA'. In the medium of refractive index n, its maximum half-angle at A is equal to 90 degrees, if D is unlimited. But the light is totally reflected at D for angles of incidence greater than a limiting angle L:

$$\sin L = \frac{1}{n} < 1. \tag{36}$$

Only a conical beam with a half-angle L or smaller coming from A may penetrate the image medium of index equal to unity. Let the wavelength λ in the medium with index of refraction n be the unit of length. For a beam from object A,

$$u^2 + v^2 = U_0^2 < 1. \tag{37}$$

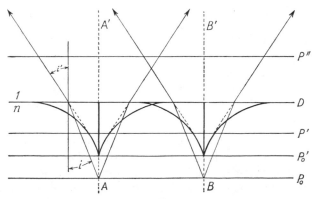

Fig. 77.

For a beam having crossed D, the condition is

$$u'^2 + v'^2 = U_0'^2 \leqslant \frac{n^2}{1} < 1. \qquad (38)$$

The two spatial frequency ranges are shown in Fig. 78 as they relate to the object space, and therefore to the information contained in the object, for the case of Fig. 77 with $n = 1.60$.

But where are the images? The images reflected on D, which are stigmatic in geometrical optics, will be left aside. They correspond to the area between the circles of radius U_0 and the circles of radius U_0' in Fig. 78. The images seen through the surface, which are virtual in this case, are not stigmatic even though they belong to beams that satisfy Abbe's condition. In order to obtain an approximate stigmatism yielding a sufficient and useful resemblance between the object and the image, the observer must impose certain conditions.

1. For incoherent imagery, in order to measure the distance between points A and B, the observer must reduce the angular extent of the beam entering his ocular to a narrow paraxial cone about AA', and then about BB'; he must focus on the plane P_0' or a bit above, where the cusp of the tangential caustic and what remains of the sagittal caustic yields a condensed image of revolution. It is impossible to obtain two good images at the same time, except if the two points are very close to each other.

2. With coherent illumination, if points A and B have a definite phase relation, a close resemblance will be obtained only under the same conditions mentioned above. In addition, there will be diffraction fringes, which are always more highly visible with coherent light.

But if it is required to obtain the maximum amount of information, instead of resemblance to the object, the appearance of the image is

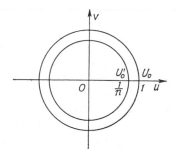

Fig. 78.

immaterial. The virtual planes like P_0' and P' and the real planes like P'' contain all the information. In particular, the nonlocalized fringes which may be observed above the diopter D allow the reconstruction by mathematical or physical means of all the information propagated in the cone of light.

The following general remark will conclude this simple example: *For a reversible beam, Abbe's condition, which is identical to the condition of isoplanatism, assures the transmission of the information between the object and the image, but not the resemblance between the object and the image.* We shall call such images *abbeian* images. It is easy to see that any section of a perfectly coherent beam from an object is an abbeian image or object: the information always remains completely available all along the beam. This is not the case for incoherent imagery: if an incoherent object is projected, the pupil of projection receives the transmitted beam according to some law that is the same for all the points of the object; only infinitesimal residues of information are available. It is this constant transformation of information in coherent beams that allows the recording of holograms and their reconstruction of images.

Abbe's condition is thus the definitive confirmation of the geometrical and trigonometric meaning of the spatial frequency variables u and v. If the wavelength λ is taken as the unit of length, u and v are the direction cosines that determine the direction of propagation OM of the diffracted light. The absolute condition

$$u^2 + v^2 \leqslant 1 \tag{39}$$

must be satisfied; and if the unit of length is not λ,

$$u^2 + v^2 \leqslant \left(\frac{1}{\lambda}\right)^2, \tag{40}$$

with

$$\frac{1}{\lambda} = \mu. \tag{41}$$

The Fourier transform $f(u, v)$ occupies a limited range of the circle of extension with the radius $\mu = 1/\lambda$. It is therefore not possible to follow the indications of Fourier theory and to define *a priori* a limited pupil. It is the spectrum $f(u, v)$ that defines $F(x, y)$ and not $F(x, y)$ that defines $f(u, v)$. F is defined by wave theory in a way that amounts to an analog calculation, and as we shall see, $f(u, v)$ is defined according to physical realities. $f(u, v)$

is a Euclidian function compatible with those realities within its extension φ, whereas $F(x)$ is necessarily a function diffracted by the internal correlation function $\Phi(x, y)$.

The variable z and its conjugate w, which is also a direction cosine, have been missing from these diffraction calculations. The excuse is immediate: the equation

$$u^2 + v^2 + w^2 = 1 \text{ or } \mu \tag{42}$$

clearly shows that of the three variables u, v, and w, only two variables are independent. All the functions are therefore two-dimensional, leading to the easy approach of plane distributions. However, the phenomena are three-dimensional, and holograms certainly show that beams of light carry three-dimensional information. Perhaps the study of three-dimensional diffraction of x-rays by crystals will clear up this dilemma.

X-Ray Diffraction

We think that the simplest and the most general form of the laws of optics with regard to image formation is to be found in x-ray diffraction. The material diffracting object is three-dimensional, which is normal; the index of refraction is very close to unity, and the absorption is very weak. It may be assumed without major error that the photons propagate among the atoms as they do in a vacuum. For a small number of photons, diffraction is a minor accident on their trajectory along a straight line, a simple and easily understood interaction between matter and radiation. On the other hand, the beam of photons is dense enough with respect to the density of atoms so that the wave picture for the radiation may be used.

Only the case of a perfectly periodic crystal will be considered. To avoid unnecessary physical hypotheses, matter is represented by a three-dimensional periodic function, which is independent of the incident radiation, and which we call the *shape function*:

$$\text{shape function} = F_0(x, y, z). \tag{43}$$

The monochromatic, coherent, and collimated incident radiation may be represented by the product of two exponentials, one a function of time, and the other of the spatial coordinates. If u_0, v_0, and w_0 are the direction cosines of the propagation vector

$$e^{i\omega t} \exp[i 2\pi(u_0 x + v_0 y + w_0 z)]; \tag{44}$$

the coherent object, whose diffraction will allow the integration of the image, is the product of the two functions defined in Eq. (43) and Eq. (44):

$$F(x, y, z) = F_0(x, y, z) \exp[i\, 2(u_0 x + v_0 y + w_0 z)]. \qquad (45)$$

The unit of length is again taken to be equal to the wavelength λ. The function $F(x, y, z)$ has the dimensions of a complex amplitude, the time dependence $\exp(i\omega t)$ common to both sides of the equation having canceled out. The volume of matter that diffracts the radiation is a secondary source of the type defined by Huyghens's principle, but in three dimensions. We call the function of Eq. (43) the *Huyghens function* of the object, and the function of Eq. (44) the Huyghens function of the radiation. Once again the time dependence has been eliminated, and the wave theory has been used to obtain the equation and to justify the rule of diffraction.

This rule is very simple: it consists of calculations similar to those that were done at the beginning of this chapter. The complex amplitude diffracted at infinity in the direction u', v', w' is given by the function

$$f(u', v', w'), \qquad (46)$$

which we call the diffraction function, and which is the Fourier transform of $F(x, y, z)$:

$$f(u', v', w') = *T[F(x, y, z)]. \qquad (47)$$

The problem is to find the shape function $F_0(x, y, z)$. Its Fourier transform $f_0(u, v, w)$ may be obtained from $f(u', v', w')$ by means of a shift in the origin:

$$u' = u + u_0,$$

$$v' = v + v_0, \qquad (48)$$

$$w' = w + w_0;$$

$$f(u', v', w') = f(u + u_0, v + v_0, w + w_0) = f_0(u, v, w). \qquad (49)$$

Ever since Laue and Ewald, this set of relations has had the following geometrical interpretation. The functions of x, y, and z are represented with cartesian coordinates in the physical space of x, y, and z ($\mathcal{R}$-space), and their transforms are represented in the Fourier space with coordinates u, v, w, with the two systems of coordinates superimposed. This is illustrated in Fig. 79.

In particular, if $F(x, y, z)$ is a periodic function, $f(u, v, w)$ degenerates into a set of integrable points at the nodes of the reciprocal lattice of the crystal lattice. The two lattices are three-dimensional. In consequence certain functions degenerate into series which are of course limited in Fourier space. In real space, the integrations are therefore reduced to the volume of a periodic lattice. On the other hand the series in Fourier space are limited to the frequency $1/\lambda$.

In Fig. 79 we assumed that the uv plane was a discrete plane and that the u-axis was a discrete line. The direction cosines u_0, v_0, w_0 are the components of a vector OO' which, given the signs of the exponents in the transformation equations, is directed oppositely to the propagation of the radiation from O' to O.

The projections of the vector PO' on the axes are u', v', and w'. It has the same length as OO', it is directed oppositely to the diffracted radiation, which goes from O' towards P. If the unit of length in physical space $\mathcal{R}$ is equal to λ, then OO' and PO' are unit vectors. If an arbitrary unit of length is used instead, then the two vectors have lengths equal to μ, and the directions cosines become λu, λv and λw.

Mathematically speaking, if physical space and the functions $F(x, y, z)$ and similar functions are postulated, the variables u, v, w are defined by the

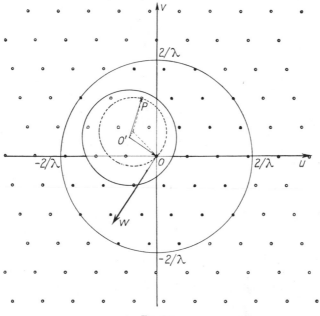

Fig. 79.

conjugate transformation equations as three independent variables over an infinite range. Physically, the vector with components u, v, w is the resultant of two vectors with constant length μ:

$$u = u' - u_0, \qquad v = v' - v_0, \qquad w = w' - w_0,$$

$$\vec{u}_0 + \vec{v}_0 + \vec{w}_0 = \mu, \qquad \vec{u}' + \vec{v}' + \vec{w}' = \mu. \tag{50}$$

The points P of Fourier space correspond to a tangible physical phenomenon, that is, the directions of diffracted radiation, which may be integrated into an image on a sphere with its center O' at a distance μ from the origin O. This is called Laue's sphere of extension. *From a single image, for one direction of illumination, only one section of the transform of the object may be determined on the sphere of extension.* In Fig. 79 only one term of the series transform has been shown on the sphere of extension. This corresponds to a well-known case. The angle $OO'P$ is equal to twice the Bragg angle.

By rotating the direction OO' about the point O, the sphere of extension can sweep the interior of a sphere having a radius equal to 2μ. This is the part of the transform of $F_0(x, y, z)$ that is accessible with radiation of wavelength λ. We call this sphere the sphere of resolution. Its volume increases as μ^3, which explains why small wavelengths are so desirable.

Two-Dimensional Distributions

By interpreting Abbe's condition as a propagation law, we have virtually reestablished the third dimension z and its dependent conjugate variable w. It seems to the author that the way out of this difficulty resides in the physical units of the complex amplitude densities. If there is propagation, we are in three-dimensional space and these densities are volume densities. We shall have to write the density at point (x, y) of the plane xOy in the form $F(x, y, z)$. The delta function will prove useful for this.

Let $\delta(z, 0)$ be Dirac's function relative to the z-axis, located at the origin of coordinates x, y. The convolution

$$F(x, y) \otimes \delta(z, 0) = F(x, y, 0), \tag{51}$$

preserves the definition of $F(x, y)$ and its numerical values, but changes its units and gives it the meaning of a volume density.

Its Fourier transform $f(u, v)$ is therefore multiplied by $H(0, w)$, the transform of $\delta(z, 0)$, which is a unit density along the w-axis:

$$*T[F(x, y) \otimes \delta(z, 0)] = f(u, v)H(0, w). \tag{52}$$

The product on the right side of Eq. (52) substitutes for the plane distribution $f(u, v)$ a cylindrical distribution with the axis of the cylinder parallel to the w-axis. This cylindrical distribution may be visualized as resulting from scanning the Fourier space with $f(u, v)$ along the w-axis from $-\infty$ to $+\infty$, as shown in Fig. 80. Because we have used the three-dimensional notation $F(x, y, 0)$, we shall write its Fourier transform $f(u, v, \underline{\infty})$, underlining the ∞ sign to indicate that it is not a value of w, but a range:

$$*T[F(x, y, 0)] = f(u, v, \underline{\infty}). \tag{53}$$

The sphere of extension is thus reestablished for those two functions, but because we are provisionally considering only light propagating through the xy plane in the $+z$ direction, the sphere is reduced to a hemisphere whose center is of necessity on the side of negative z, as illustrated in Fig. 80.

Figure 80 shows the plane uOw and its intersection with the sphere. The center O' of the sphere of extension of radius μ is placed in the plane of Fig. 80 at the point $u_0, v_0 = 0, w_0$. The light is incident from below. It crosses the xy plane (which is assumed coincident with the uv plane in two-dimensional Fourier space) in the direction of increasing w; w_0 is therefore negative. The incident light has direction $O'O$, and $O'P$ is a direction of diffracted light, also crossing the xy plane in the direction of increasing w or z. Draw a straight line through P parallel to the z, w axis; this is the intersection of $H(0, w)$ with the plane at a point u, v corresponding to P.

It is easy to see that the surface integrals along the sphere of extension are reduced in this case (as well as for x-rays) to plane integrals in uv. They are projections in the elementary geometrical sense, which neglects physical realities and dimensions.

Figure 80 illustrates a type of illumination that many techniques have in common: microscopy, spectroscopy, interferometry, reflection, refraction,

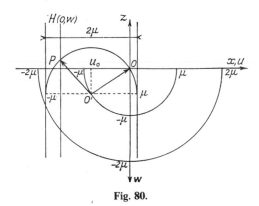

Fig. 80.

and polarization. O' may, at least in principle, represent the hemisphere having a radius μ and a center O. The spheres of extension that have their centers on this hemisphere are tangent to the lower half of the sphere of resolution. The projection of the latter in the uv plane yields the circle of resolution of radius 2μ. Abbe has shown that this sphere limits the range of explorable frequencies. This is an absolute limitation, which unfortunately belongs to the murky realm of far-field diffraction.

Figure 81 shows three classical positions of the circle of extension inside the circle of resolution:

$$\text{Circles of extension:} \quad \begin{cases} \text{center } O & \text{normal incidence,} \\ \text{center } O' & \text{oblique incidence,} \\ \text{center } O'' & \text{grazing incidence.} \end{cases}$$

From now on, we abandon the simultaneous representation of the x, y, z and the u, v, w axes as superimposed axes. In Fig. 80 this convention forced us to represent the positive direction of w along the negative direction of z. For clarity and mathematical simplicity, it is better to represent the axes of physical space and Fourier space independently of each other, by means of ordinary cartesian axes. The signs may be easily obtained from the disposition of the images. From now on, in the three-dimensional space u, v, w, a unit vector or a vector of length μ will represent the direction of propagation of the light, as illustrated in Fig. 82.

The Two Fourier Transforms

The sphere with a radius of unity or $\mu = 1/\lambda$ is what we call the *sphere of radiation*; its radii are Euclidian directions of light propagation. The plane

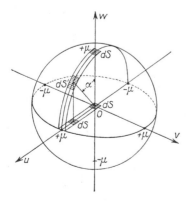

Fig. 81.

functions $F(x, y)$ and $F(x, y, 0)$, two equivalent ways of representing the same complex amplitude distribution, each have a Fourier transform: one is the plane distribution $f(u, v)$, and the other is spread over the sphere of radiation, and is the intersection of the sphere with the cylindrical distribution $f(u, v, \infty)$, at least for its positive part, which we write $f(u, v, w)$, w being positive. A relation between $f(u, v)$ and $f(u, v, w)$ may be established by replacing $f(u, v)$ with the range function $\varphi(u, v)$, whose density is equal to unity. Let $\varphi'(u, v, w)$ be the corresponding distribution on the hemisphere of extension. It is obvious that the integration of the latter with respect to w is a normal projection on the uv plane:

$$\int_\varphi \varphi'(u, v, w) \, dw = \varphi(u, v). \qquad (54)$$

The hemispherical range function therefore does not have a constant density. If the angle between OM and Ow is α, the relation between the element ds' centered on M and its projection $ds = du \, dv$ is

$$\varphi'(u, v, w) \, ds' = \varphi(u, v) \, ds \qquad (55)$$

which means that

$$\varphi'(u, v, w) = \varphi(uv)\cos \alpha = w\varphi(u, v)$$

$$= w, \qquad (56)$$

because $\varphi(u, v)$ is equal to unity.

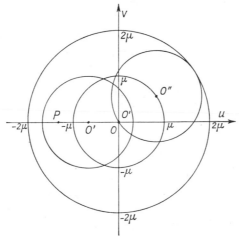

Fig. 82.

Lambert's Law

The last expression is reminiscent of Lambert's law, by virtue of the factor $\cos \alpha$. We recall its classical formulation. Let an element of area $dS = dx\,dy$ have a constant luminance L which is independent of the mean direction α of the cone of radiation having a solid angle of radiation $d\omega$ measurable on a sphere with a radius equal to unity. At a distance r from dS, $d\Sigma$ is an element of area illuminated at an angle of incidence θ by the cone $d\omega$. The flux from the source dS on the area $d\Sigma$ is equal to

$$d\Psi = \frac{dS\,d\Sigma \cos \alpha \cos \theta}{r^2}L. \tag{57}$$

The classical element $d\omega$ is the element dS of Eq. (55). The element $d\Sigma$ is at infinity and $L = 1$ for plane range functions. There remains

$$d\Psi = dS\,ds\,w. \tag{58}$$

Lambert's law is used in photometry and in illumination engineering for scalar energy distributions that add together by superposition. Here it is applied to complex amplitude distributions that add like vectors. But in both cases the differential law is the same.

Range and Internal Correlation Functions for One, Two, and Three Variables

We now consider functions whose most extended range functions are limited by the wavelength. As usual, let $\mu = 1/\lambda$.

Range of a Single Variable.   The range function is a rectangle function with a value equal to one over the range $-\mu$ to $+\mu$:

$$\Phi(x) = T[\varphi(u)] = 2\mu\frac{\sin 2\pi\mu x}{2\pi\mu x}. \tag{59}$$

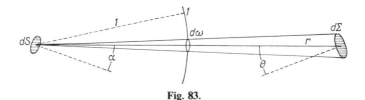

Fig. 83.

If λ and μ are set equal to unity, then

$$\Phi_0(x) = 2\frac{\sin 2\pi x}{2\pi x}. \tag{60}$$

Range of Three Variables. The basic range function is a layer of unit density $\sigma(u, v, w)$ on the surface of the sphere of radiation with radius μ. Its Fourier transform is equal to $\Sigma(x, y, z)$. These two functions have spherical symmetry, and it is sufficient to have their values along a diameter:

$$\Sigma(x, 0, 0) = T\left[\int\int_{-\infty}^{+\infty}\sigma(u, v, w) \, dv \, dw\right].$$

This integration yields a function of u that is a rectangle function equal to $2\pi\mu$, with a range 2μ along the x-axis; except for a factor of 2π, this is equal to the function of Eq. 60:

$$\Sigma_0(x, 0, 0) = 4\pi\frac{\sin 2\pi x}{2\pi x}. \tag{62}$$

Because of the spherical symmetry of the functions σ and Σ, this distribution of three variables is usually represented by a function of the single variable r:

$$\Sigma_0(r) = 4\pi\frac{\sin 2\pi r}{2\pi r}. \tag{63}$$

The complex amplitude distribution given by this function is identical, except for a constant factor, to the distribution in a system of stationary spherical fringes produced by a point source at the focus of a suitable complete spherical mirror.

Range of Two Variables. We have found two equivalent equations for the maximum range function:

a. For a plane transform, $\varphi(u, v) = 1$ inside the circle with radius μ.
b. For a spherical transform, $\varphi'(u, v, w) = w$ on the positive hemisphere with radius μ.

Consider them one after the other.

a. The function $\varphi(u, v)$ has circular symmetry, along with its Fourier transform $\Phi(x, y)$; the latter is the Airy diffraction pattern, which is completely determined by values along one radius. If $\lambda = 1$, then

$$\Phi_0(x,0) = \pi \frac{2J_1 2\pi x}{2\pi x} .$$

This may be written as a function of the radius ρ from zero to infinity in the form

$$\Phi_0(\rho) = \pi \frac{J_1 2\pi \rho}{\pi \rho} .$$

b.   There are two equivalent ways to integrate the transform $\Phi(x, y, 0)$ of $\varphi'(u, v, w)$. The first is to integrate with respect to w, z being equal to zero:

$$\Phi(x, y, 0) = \int\int\int_{-\infty}^{+\infty} \varphi'(u, v, w) \exp(ux + vy) \, du \, dv \, dw. \qquad (66)$$

The integration with respect to w projects φ' on the uv plane, which yields the two-dimensional transform

$$\Phi_0(x, y) = \int\int_{-\infty}^{+\infty} \varphi(u, v) \exp(ux + vy) \, du \, dv, \qquad (67)$$

which leads to Eq. (65). But Eq. (56) may be written

$$\varphi'(u, v, w) = \sigma(u, v, w) \cdot w \qquad \text{for} \quad 0 < w < 1. \qquad (68)$$

In Chapter 3, in Eq. (9) and Fig. 35b, were given the tools required to calculate the transform of the function

$$g(w) = w \qquad \text{for} \quad 0 < w < 1, \qquad (69)$$

namely,

$$G(z) = \frac{\sin 2\pi z}{2\pi z} - \frac{1 - \cos 2\pi z}{(2\pi z)^2}, \qquad \text{with} \quad z \geqslant 0. \qquad (70)$$

$\Phi_0(\rho)$ is given by the convolution

$$\Phi_0(\rho) = \Phi_0(r) \otimes G(z). \qquad (71)$$

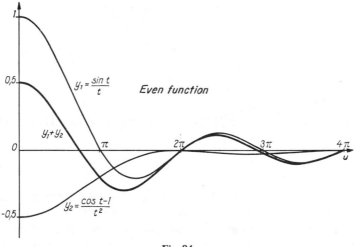

Fig. 84.

The three internal correlation functions are thus related to the three-dimensional correlation function. A graph of $G(z)$ and its two components is shown in Fig. 84; the cosine component disappears very fast.

Thus the convolution of Eq. (71) is a particular mode of projection for the case of interest here.

Note. Lambert's law and Figs. 82 and 83 are as valid for incoherent imaging as they are for coherent imaging. The distribution of energy on the sphere of radius μ is the spatial distribution of intensity that was established for finite sources. The energy distribution is obtained from the distribution of the complex amplitude by multiplying $f(u, v, w)$ by its complex conjugate $f^*(u, v, w)$. We shall not emphasize further this well-known aspect of directional spheres.

7

Plancherel's Theorem and Correlation

INTRODUCTION

In the first edition of this book, the title of Chapter 7 was almost identical to this one. Incoherent imaging was the fundamental type of imaging, the kind that one usually observed. It had been stretched beyond the limits of human perception into the long and short wavelengths, but it was for the narrow range of visible frequencies that it related to perceptual reality. The image perceived by the eye or recorded by a photographic emulsion is one of energy distribution; coherent imaging is connected with this detection process through Plancherel's theorem.

We shall see in the following chapters that the relations between coherence and incoherence are complex. In Chapter 8, the phenomenon of interference by translated objects, together with one of Wood's stimulating remarks, will force us to see a point as the perfect source of coherence, as in Fresnel's time; in Chapter 9 this will be extended to "information points." The application of Dirichlet's theorem to incoherent images will lead to the definition of two types of internal correlation for those images: one, derived directly from Dirichlet's theorem, is actually cumbersome and useless; the other, founded on the individual coherence of points, is useful, but yields for the internal correlation the square of the modulus of the correlation for coherent imaging.

Those difficulties will be dealt with as they arise.

THE FIRST FORM OF PARSEVAL'S THEOREM

Product of Two Functions

Let two functions $F(x)$ and $G(x)$ have the Fourier transforms $f(u)$ and $g(u)$:

$$F(x) = T[f(u)],$$
$$G(x) = T[g(u)]. \tag{1}$$

The Fourier transform of the product $F(x)G(x)$ is the convolution of their transforms:

$$\int_{-\infty}^{+\infty}[F(x)G(x)]\text{cis}(-2\pi ux)\,dx = \int_{-\infty}^{+\infty}f(u')g(u-u')\,du'. \quad (2)$$

Let $u = 0$. Dropping the useless prime on u', Eq. (2) becomes

$$\int_{-\infty}^{+\infty}F(x)G(x)\,dx = \int_{-\infty}^{+\infty}f(u)g(-u)\,du = \int_{-\infty}^{+\infty}f(-u)g(u)\,du. \quad (3)$$

If instead of integrating the product $F(x)G(x)$, we integrated the product of their transforms, the reciprocal relation would be

$$\int_{-\infty}^{+\infty}f(u)g(u)\,du = \int_{-\infty}^{+\infty}F(x)G(-x)\,dx = \int_{-\infty}^{+\infty}F(-x)G(x)\,dx. \quad (4)$$

The Square of a Function

If in Eq. (2) $F(x)$ is equal to $G(x)$, then

$$\int_{-\infty}^{+\infty}F^2(x)\,dx = \int_{-\infty}^{+\infty}f(u)f(-u)\,du. \quad (5)$$

The spectrum of $F^2(x)$ is obviously equal to

$$*T[F^2(x)] = f(u) \otimes f(-u) = j(u). \quad (6)$$

The reciprocal relations involving the conjugate functions are

$$F(x) \otimes F(-x) = T[f^2(u)], \quad (7)$$

$$\int_{-\infty}^{+\infty}F(x)F(-x)\,dx = \int_{-\infty}^{+\infty}f^2(u)\,du. \quad (8)$$

The square of a complex function is itself a complex function:

$$(\rho e^{i\varphi})^2 = \rho^2 e^{i2\varphi}. \quad (9)$$

The terms on both sides of this expression contain the same information.

PLANCHEREL'S THEOREM

This theorem complements those above. In Eq. (3), replace $G(x)$ and $g(u)$ by $F^*(x)$ and its transform $f_j(u)$:

$$\int_{-\infty}^{+\infty} F(x) F^*(x)\, dx = \int_{-\infty}^{+\infty} f(u) f_j(-u)\, du. \tag{10}$$

From Eq. (2.47),

$$f_j(-u) = f^*(u); \tag{11}$$

so Eq. (10) may be written

$$\int_{-\infty}^{+\infty} F(x) F^*(x) = \int_{-\infty}^{+\infty} f(u) f^*(u)\, du, \tag{12}$$

$$\int_{-\infty}^{+\infty} |F(x)|^2\, dx = \int_{-\infty}^{+\infty} |f(u)|^2\, du. \tag{13}$$

The integral of the modulus squared of a function is equal to the integral of the modulus squared of its Fourier transform.

It is easy to derive from the preceding considerations the Fourier transform of the squared modulus of $F(x)$:

$$*T[F^*(x)] = f_j(u) = f^*(-u), \tag{14}$$

$$*T[F(x) F^*(u)] = f(u) \otimes f^*(-u) = d(u). \tag{15}$$

The square of the modulus of a complex function is a real function. As a parallel to Eq. (9), we have

$$\left| \rho e^{i\varphi} \right|^2 = \rho^2. \tag{16}$$

The phase is gone, and there is a considerable loss of information between the functions on either side of the equation. Holography has taught us that it is the depth information that is lost. In practice the equation is not reversible.

The function $d(u)$ plays a considerable role in any imaging theory. It is related to the correlation functions, about which a few words will be said at the end of this chapter. It is called the *autocorrelation function* of $f(u)$.

The mathematical theory of correlations touches optics and its Fourier transforms, but the expressions correlation function and autocorrelation function are rather vague in physics, and we have been led to give the former name to the transforms of range functions, for want of anything better. In physics everything is a correlation function in the wider sense of the word, except perhaps for the neutrino flux, for the time being.

In coherent imaging the phase remains available all along the beam. To extract it is obviously more difficult than looking in an ocular, but holography has shown that it is possible. For each point in a plane section across the beam, the weighting functions are complex amplitude distributions that interfere without losing their individuality, whatever their behavior.

For incoherent imaging, the phase loses its additivity for the weighting functions of neighboring points, which are in fact, starting right at the object, extended neighboring elements that add like distributions of energy. They also change all along the beam. Consider the simple case of a self-luminous object from which an image is formed with a convergent lens. At the lens, the sections of the coherent beams from the object points add together to form a uniform distribution of intensity which contains all the available energy. In order to obtain a resemblance between the object and the image, albeit with some loss in information, a maximum concentration of the weighting functions—that is, a focusing—must be found.

APPLICATION TO FOURIER SERIES

Of course Eq. (13) is also valid for Fourier series and for functions of more than one variable. For the various forms of series that we have introduced, the appropriate expressions with their corresponding equation numbers in the preceding chapters are

$$\int_{-\infty}^{+\infty} |F(x)|^2 = \left[\frac{a_0}{2}\right]^2 + \sum_{n=1}^{+\infty} \left(a_n^2 + b_n^2\right), \tag{1.10}$$

$$\int_{-\infty}^{+\infty} |F(x)|^2 = \left[\frac{a_0}{2}\right]^2 + \sum_{n=1}^{+\infty} \mathcal{C}_n^2, \tag{1.37}$$

$$F(x)^2 = \sum_{n=-\infty}^{+\infty} |Z_n|^2. \tag{2.28}$$

For series as well as for integrals, the differences between the functions on both sides of the equations is a measure of the errors that result from the operations of squaring and summation.

THE CONSERVATION OF ENERGY

The distribution of energy corresponding to a complex amplitude distribution $F(x)$ is the distribution of the square of the modulus $|F(x)|^2$. Equation (13) is evidently an expression for the conservation of energy. Its normal form is two-dimensional:

$$\iint_{-\infty}^{+\infty} |F(x, y)|^2 \, dx \, dy = \iint_\varphi |f(u, v)|^2 \, du \, dv. \tag{17}$$

Even if it were not derived from the Fourier transform, it would have exactly the same form. It is a consequence of the linearity of the propagation from the pupil to the diffraction pattern "at infinity." But from the point of view of physical reality, this is not without difficulties, mostly due to the fact that the range of $f(u)$ or $f(u, v)$ is always limited by the circle of extension or by the sphere of illumination.

For monochromatic light, all the photons carry the same amount of energy w and both terms of the following equation contain the same number N:

$$Nw = W_0 = \iint_\varphi |f(u, v)|^2 \, du \, dv. \tag{18}$$

The function $|f(u, v)|^2$ remains Euclidian with the same range as $f(u, v)$. It is only when they are dispersed at infinity that the isolated photons, having lost all possibility of interaction with each other and in consequence any possible phase relation, end up being countable. But in reality it is sufficient that they be resolvable on a sphere with a great enough radius. Experience shows that the photon energy is available only inside a sphere with a radius smaller than a single wavelength. At this distance the photon energies are resolvable. Their number is finite, and the continuous dispersion

$$W_0(u, v) = |f(u, v)|^2 \tag{19}$$

becomes a continuous probability distribution, as was the case for $F(x)$.

GRAPHICAL REPRESENTATIONS

Decompositions of Complex Integrals

Convolution integrals such as the ones on the right sides of Eqs. (2), (6), and (15), where the two functions $f(u)$ and $g(u)$ are complex, may be reduced to convolutions between real functions. Separate the real and imaginary parts in the usual way:

$$f = f' + if'', \qquad g = g' + ig''. \tag{20}$$

The general expression (2) becomes

$$f \otimes g = (f' \otimes g') - (f'' \otimes g'') + i[(f' \otimes g'') + (f'' \otimes g')], \tag{21}$$

an expression that has four independent convolutions. In Eq. (6) for the squared function, $g(u)$ is replaced by $f(u)$:

$$f \otimes f = (f' \otimes f') - (f'' \otimes f'') + 2i(f' \otimes f''), \tag{22}$$

which contains three distinct real integrals. In the autocorrelation equation (1), $g(u)$ is replaced by $f^*(-u)$, which we represent by $f^*(-)$, and which is equal to $f'(-u) - if''(-u)$:

$$f \otimes f^*(-) = f' \otimes f'(-) + f'' \otimes f''(-) - i[f' \otimes f''(-) - f'' \otimes f'(-)]. \tag{23}$$

This is the function $d(u)$ of Eq. (15). Like the function of Eq. (21), it depends *a priori* on four real convolutions. However, the result is simpler than that of Eq. (22): the function $d(u)$ is the square of the Fourier transform of a real and positive function—the square of a modulus. It consists of a real even part, which is the transform of the even part of the squared function, and an odd imaginary part, which is the transform of the odd part of the squared function. One may expect to find simplifications by considering parity.

When f and g are real functions in Eq. (21), things are really simple. There is only one integration on one variable, which is easy to represent graphically.

Parity. Equations (22) and (23) present two types of real and simple convolutions

$$f(u) \otimes f(u) \quad \text{and} \quad f(u) \otimes f(-u),$$

which may be written in abridged form as $f \otimes f$ and $f \otimes f(-)$:

$$f \otimes f = (f_e + f_o) \otimes (f_e + f_o),$$

$$= (f_e \otimes f_e) + (f_e \otimes f_o) + (f_o \otimes f_e) + (f_o \otimes f_o),$$

$$= (f_e \otimes f_e) + (f_o \otimes f_o) + 2(f_e \otimes f_o). \tag{24}$$

The two functions in the first convolution do not change signs when u is changed to $-u$: the convolution is even;

the two functions of the second convolution both change signs when u is changed to $-u$: this convolution is odd;

only one of the two functions of the third convolution changes its sign when u is changed to $-u$: this convolution is odd.

Hence

$$f \otimes f(-) = (f_e + f_o) \otimes (f_e - f_o),$$

$$= (f_e \otimes f_e) - (f_e \otimes f_o) + (f_o \otimes f_e) - (f_o \otimes f_o),$$

$$= (f_e \otimes f_e) - (f_o \otimes f_o). \tag{25}$$

The real part of the second member of Eq. (23), which contains two convolutions of this type, is therefore even. It remains only to verify that the quantity in brackets at the end of Eq. (23) is indeed odd:

$$f' \otimes f''(-) - f'' \otimes f'(-)$$

$$= (f'_e + f'_o) \otimes (f''_e - f''_o) - (f''_e + f''_o) \otimes (f'_e - f'_o)$$

$$= \left[\begin{array}{c} (f'_e \otimes f''_e) - (f'_e \otimes f''_o) \\ + (f'_o \otimes f''_e) - (f'_o \otimes f''_o) \end{array} \right] - \left[\begin{array}{c} (f''_e \otimes f'_e) - (f''_e \otimes f'_o) \\ + (f''_o \otimes f'_e) - (f''_o \otimes f'_o) \end{array} \right]$$

$$= -2 \left[(f'_e \otimes f''_o) - (f'_o \otimes f''_c) \right]. \tag{26}$$

The simplification of Eq. (23) is thus

$$f \otimes f^*(-) = (f'_e \otimes f'_e) - (f'_o \otimes f'_o) + (f''_e \otimes f''_e) - (f''_o \otimes f''_o)$$

$$+ 2i \left[(f'_e \otimes f''_o) - (f'_o \otimes f''_e) \right]. \tag{27}$$

There remain only six convolutions out of the original twelve; this is a manifestation of the loss of spectral information related to the dropping of the phase information when the modulus is squared.

Note. If the signs of the variables are changed, that is, if $f(u)$ is replaced by $f(-u)$, the parity of the function does not change with respect to the variable or with respect to the variables with changed signs.

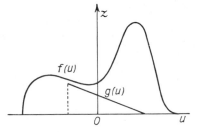

Fig. 85.

Volume of Integration of Quadratic Spectra

Ordinary Convolution.   Let two continuous limited functions $f(u)$ and $g(u)$ be represented by two plane and real curves as shown in Fig. 85. Consider a system of cartesian axes Os, Ot, Oz, where the coordinates s and t have the same scale as the coordinate u. In Fig. 86, the function $f(u)$ is drawn in the plane sOz, and $g(u)$ is drawn in the plane tOz. Let

$$z = f(s) = f(u) \quad \text{and} \quad z' = g(t) = g(u). \tag{28}$$

Let a plane $t'Oz'$ parallel to tOz be translated along the s axis. The plane tOz intersects $f(s)$ at the point A with $OA = f(0)$, which may be taken as the unit of length along z. On the translated plane that intersects $f(s)$ at A', draw the function

$$g'(t') = \frac{OA'}{OA} g(t) = f(s)g(t) = \frac{1}{f(0)} f(u)g(u). \tag{29}$$

As the translated plane slides along the range of $f(s)$, this curve generates a volume $f(s)g(t)$, which we call the volume of integration of the convolution.

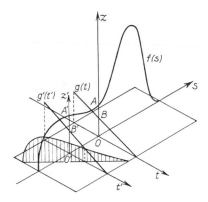

Fig. 86.

Let this volume be intersected by the vertical plane

$$s + t = u = \text{constant.} \tag{30}$$

The area of this section (not represented in Fig. 86) is

$$\int f(s) g(u - s) \, ds \sqrt{2} = \sqrt{2} \, j(u), \tag{31}$$

$$j(u) = f(u) \otimes g(u). \tag{32}$$

The factor $\sqrt{2}$ comes from the stretching of the variable of integration along the base of the oblique section.

The intersection of the volume by the plane

$$s - t = u = \text{constant} \tag{33}$$

is shown in Fig. 86 as the area with vertical hatching. This area is equal to

$$\sqrt{2} \int_{-\infty}^{+\infty} f(s) g(s - u) \, ds = \sqrt{2} \, d(u), \tag{34}$$

$$d(u) = f(u) \otimes g(-u). \tag{35}$$

Volume of Integration of Squared Functions

It is easy to apply the foregoing analysis to the two kinds of integration that we considered in the section on parity. If $g(u) = f(u) f(u)$ over a limited range, the surface of integration is a square. The plane shown in Fig. 87 represents a linear function with an even component and an odd compo-

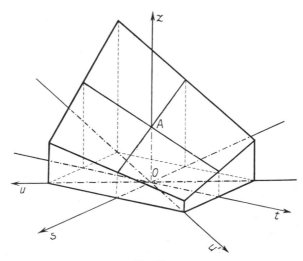

Fig. 87.

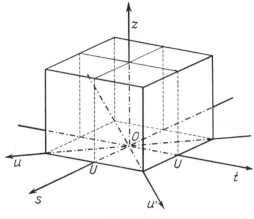

Fig. 88.

nent, seen from the positive quadrant of s, t. The straight line Ou is the axis of convolution of Eq. (32), and the straight line Ou' is the axis of convolution of Eq. (35). These straight lines bisect the angles between the axes. If the extremities of the functions along the s- and t-axes are at a distance U from the origin O, the two diagonals must be scaled so that the extremities of the convolutions are at a distance $2U$ from the origin O.

Amplitude Spectrum of a Slit. Let $f(u)$ be equal to a constant in the range $-U$ to $+U$, and zero outside. The volume of integration is a right-angled parallelepiped with a square base shown in Fig. 88. Let its height be equal to unity. The square of the function $f(u)$ is equal to the square of its modulus; its autocorrelation is represented in Fig. 89 as an isosceles triangle with a base equal to $4U$ and a height equal to $4U$.

Amplitude Spectrum of Two Parallel Slits. Let $f(u)$ be equal to a constant inside the two slits shown in Fig. 90. The volume of integration consists of

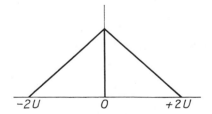

Fig. 89.

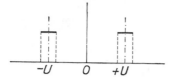

Fig. 90.

the four right-angled parallelepipeds with square bases shown in Fig. 91.
The two slits are shown along the s- and t-axes.

The quadratic spectrum is shown in Fig. (92). The central triangle is the
frequency transfer function of one slit, and the two outer triangles are the
transfer functions of the energy of the interference fringes.

The maximum frequencies $-U$ and $+U$ of Fig. 90 and $-2U$ and $+2U$
are indicated in the figures. It is sometimes useful to consider the two slits
as derived from the superposition of a screen with opaque bands on an
initial slit having a width $4U$, rather than from the widening of delta
functions located at the centers of the slits.

Amplitude Spectrum of a Grating with N Lines.   Let the grating $f'(u)$ be
built up of N equally spaced delta functions with period p along the u-axis.
There are eleven delta functions in Fig. 93; the sixth coincides with the
origin.

If this grating is compared with the rectangle function having the same
total width, it is immediately seen that the Fourier transform $F'(x)$ is the
associated periodic function of the Fourier transform $F(x)$ of the continuous slit having the same width $10p$. The volume of integration is a bed of
nails with 121 delta functions in a square array, illustrated in Fig. 94.

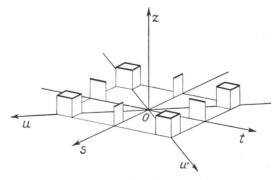

Fig. 91.

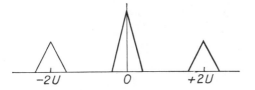

Fig. 92.

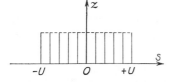

Fig. 93.

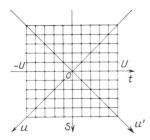

Fig. 94.

The quadratic spectrum, shown in Fig. 95, should be compared with that shown in Fig. 89; $|F'(x)|^2$ is the associated periodic function of $|F(x)|^2$.

We omit here those elementary geometrical considerations which, because of their simplicity, had a singular charm for the author while he was discovering the usefulness of the Fourier transform between 1940 and 1943.

Positions of the Spectrum of Squared Functions and the Quadratic Spectrum

The reader will forgive the choice of a function that is really too simple, but which has the advantage of being a range function and which leads to graphs composed of straight lines that may be easily drawn. The second method of image formation will be considered here.

The rectangle function $f(u)$ in Fig. 96 is real and has a value equal to one over its range AB:

$$U = u_1 - u_0. \tag{36}$$

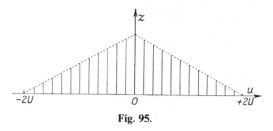

Fig. 95.

Below its graph, on a replication $O'u$ of the u-axis, is the graph of the dissipation, which is again equal to $f(u)$. For the second procedure of image formation, the function $f''(-u)$ that is the reflection of $f(u)$ in the axis Oa must be used; its support is $A''B''$ relative to an origin O'' superimposed upon the origin O. When O'' is translated along with $F''(-u)$, the convolution

$$\int f(u) f''(-u)\, du$$

may be generated over the common range of the two functions.

The convolution $j(u)$ extends from $2u_0$ to $2u_1$. The squaring of the function results in doubling the frequency range of the spectrum. If the function is odd by virtue of its position, the convolution $j(u)$ has the same parity.

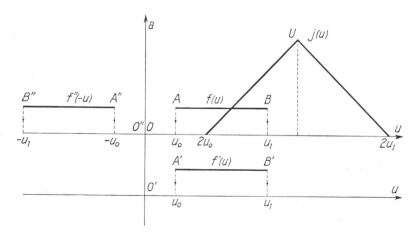

Fig. 96.

Quadratic Spectrum. Consider again Eq. (15):

$$d(u) = f(u) \otimes f^*(-u). \tag{15}$$

Consider again also Fig. 96 with its rectangle function AB. Along the replicated axis $O'u$ the dissipation function is graphed as $f(-u)$, the rectangle function with support $B'A'$, because $f(u)$ is real. By changing the sign of u in this dissipation graph according to the rules, the function $f(u)$ appears again, $A''B''$ coincides with AB, and O'' is superimposed upon O, as shown in Fig. 97a.

Let U once more be equal to the range of $f(u)$. When the origin O'' is translated along with $A''B''$, the integral that results is

$$\int f(u)f''(u) \, du,$$

once more over the range common to the two functions, that is, $2U$.

The maximum value of the correlation is found when the two axes coincide. The correlation $d(u)$ is the same triangle that was found for the square of the function, but it is now an even function about the origin, with its maximum value located at a frequency equal to zero.

The Correlation Equation. In this section we remain inside the realm of mathematics. The correlation equation has a precise meaning, particularly in holography, and the reader is referred to the numerous publications on the subject. At the beginning, correlations had the aim of measuring the degree of resemblance between two functions $F(x)$ and $G(x)$, one of the two usually being a gaussian distribution. The functions were real. The Fourier transform has allowed a more accurate evaluation of this resemblance. We have not only the functions but also their Fourier transforms

$$F(x) = T[f(u)],$$
$$G(x) = T[g(u)]. \tag{1}$$

Now the multiplication of two functions is what is called in optics a filtering operation, which may be carried out in either space:

$$F(x)G(x) = T[f(u) \otimes g(u)], \tag{37}$$

$$F(x) \otimes G(x) = T[f(u)g(u)]. \tag{38}$$

But the expression (15) for the squared modulus of a function suggests another possibility of comparison, which in holography has a precise meaning: take the complements of Eq. (37) and Eq. (38):

$$F(x)G^*(x) = T[f(u) \otimes g^*(-u)], \tag{39}$$

$$F(x) \otimes G^*(-x) = T[f(u)g^*(u)]. \tag{40}$$

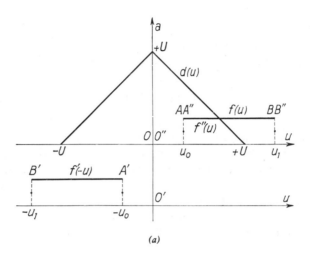

(a)

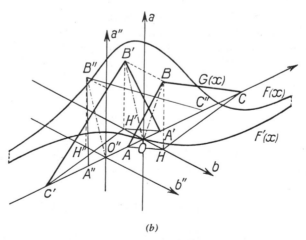

(b)

Fig. 97.

The differences between the convolution equations (37) and (38) and the correlation equations (39) and (40) are easy to see. In terms of the modulus and the phase, the functions may be written

$$F(x) = |F|\exp[i\varphi(x)],$$
$$G(x) = |G|\exp[i\psi(x)],$$

(41)

$$G^*(x) = |G|\exp[-i\psi(x)].$$

(42)

The algebraic product in Eq. (37) is

$$F(x)G(x) = |F| \cdot |G|\exp\{i[\varphi(x) + \psi(x)]\}.$$

(43)

The algebraic product of Eq. (39) is

$$F(x)G^*(x) = |F| \cdot |G|\exp\{i[\varphi(x) - \psi(x)]\}.$$

(44)

The product $|F| \cdot |G|$ has the dimensions of energy, and the distributions (43) and (44) are distributions of energy; but they are modulated by the complex exponentials that multiply them. In the convolution equations, the phases add, whereas in the correlation equations, they subtract.

The two convolutions of Eqs. (36) and (39) are illustrated in Figs. (53) and (54) of Chapter 4, and in Fig. 97b. The former figures are related to the two modes of integration of images by convolution. In Fig. 53, the product $F(x')G(x - x')$ was integrated directly. In Fig. 54, the following inversion was carried out first:

$$G'(x) = G(-x),$$

(45)

and the product $F(x')G'(x' - x)$ was then integrated. This procedure is often used. The functions $F(x)$ and $G(x)$ of Fig. 97b are the same as those of the two figures of Chapter 4, except for the additional functions

$$G^*(-x): A'B'C',$$

$$G'^*(x): A''B''C''.$$

The two dissipation graphs ABC [$g(x)$] and $A'B'C'$ [$G^*(-x)$] for the first mode of image integration are symmetrical with respect to the real axis Oa; the two graphs $A'B'C'$ (Fig. 53) and $A''B''C''$ (Fig. 97b) for the second mode of image integration are also symmetrical with respect to the real axis $O'a'$ of the shifted system of axes. This symmetry is therefore a much more characteristic property of correlation than is the sign of x.

8

Stigmatic Pupils

INTRODUCTION

The elementary and simple sources used by the optical scientist in his graphical representations have very special properties. Sources are points or sets of points, lines are infinitely thin, and apertures have perfectly smooth edges. Pupils usually consist of apertures in opaque screens. Sources and pupils are often symmetrical or periodical. Their physical definition is simple, and their introduction into the Fourier transform does not raise any special difficulty; we are always in the realm of linear optics.

In this chapter we shall abandon the preoccupations of the physicist. We accept the violation of the principle of conservation of energy caused by limiting a wavefront or a beam of photons by means of a classical diaphragm. The resulting errors are mostly inaccessible, restricted to high frequencies, lost in the noise, and far from the paraxial range, where, as we well know, the results are excellent.

We hope that the reader will not object to the use of the traditional terminology of physical optics. It is evident that the age-old distinction between interference and diffraction has no sense: in our view, linear interference and diffraction are both represented by the Fourier transform.

INTERFERENCE FRINGES

Interference by Translation

Let $F(x)$ be a distribution function of one variable, with a Fourier transform equal to $f(u)$:

$$*T[F(x)] = f(u). \tag{1}$$

By means of two translation $-h'$ and $-h''$, two other distributions $F(x + h')$ and $F(x + h'')$ may be obtained. Those distributions, illustrated in Fig. 98, have the Fourier transforms

$$*T[F(x + h')] = f(u)\text{cis}(2\pi h'u), \tag{2}$$

$$*T[F(x + h'')] = f(u)\text{cis}(2\pi h''u). \tag{3}$$

The sum of the Fourier transforms is equal to the Fourier transform of the sum:

$$*T[F(x + h') + F(x + h'')] = f(u)[\text{cis}(2\pi h'u) + \text{cis}(2\pi h''u)];$$

$$= f(u)\text{cis}\,\pi(h' + h'')u\cos\pi(h' - h'')u. \tag{4}$$

The cis factor represents a helical twist or linear phase shift which is equal to zero when $h' + h'' = 0$, that is, when the two translations h' and h'' are symmetrical with respect to the origin O and consequently with respect to $F(x)$. The cis factor does not alter the moduli of the functions.

The trigonometric cosine factor causes zeros to appear in the total transform when

$$(h' - h'')u = n + 0.5, \tag{5}$$

n a positive or negative integer. The complex amplitude fringes are shown in Fig. 99a. The observable energy distribution is found by squaring the modulus of Eq. (5). The helical factor disappears, and the frequency of the fringes is doubled, as shown in Fig. 99b:

$$E(u) = 4|f(u)|^2\cos^2\pi(h' - h'')u. \tag{6}$$

However, neither the period nor the parity of the zeros has changed; their position is determined by the translation by $h' - h''$, which superimposes $F(x + h')$ on $F(x + h'')$.

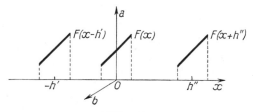

Fig. 98.

In general the function $F(x)$ is a linear distribution obtained from the linear integration of a surface distribution $F_0(x, y)$:

$$F(x) = \int_{-\infty}^{+\infty} F_0(x, y)\operatorname{cis} 2\pi v y \, dy. \tag{7}$$

The zeros are then dark fringes parallel to the y-axis, which are characteristic of interference phenomena. The amount of light passing between them depends not only on the cosine-squared factor, but also on the Fourier transform of $F(x)$ or $F_0(x, y)$. This function $F_0(x, y)$ is worth examining. Two cases may occur:

1. The two-dimensional distributions are coherent. This case has become common since the invention of the gas laser. In this case $F_0(x, y)$ and $F(x)$ are coherent, and the foregoing calculations are all valid and do not warrant further discussion.

2. The light coming from the distributions originates from an incoherent source. Let the light be perfectly incoherent at the source: because the light from two distinct points of the source does not interfere, F' and F'' are images. From our considerations of Dirichlet's theorem, all surface distributions obtained by imaging are necessarily diffracted functions whose elements are what we have called range functions, diffraction functions, or internal correlation functions. These functions are perfectly coherent, because they are derived from nonresolvable points on a surface considered as a secondary source. Let P_0 be one of those points. It is to this point that we attribute the function $F_0(x, y)$. The only condition required is that all the points similar to P_0 be replicated by the translation $h' - h''$. Each pair of images corresponding to the points P' and P'' yields the same system of

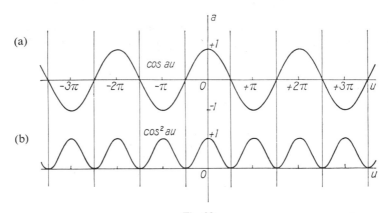

Fig. 99.

hyperbolic fringes with the axis $P'P''$. The fringes are formed at infinity, that is, beginning at a distance where the distance between the fringes becomes much greater than the dimensions of the initial distributions.

The most difficult condition to satisfy is the precision of the translation by $h' - h''$; this will be seen in the classical example of Fresnel's mirrors.

It is worth taking note of the peculiar way in which partial coherence has been analyzed in the foregoing.

Fresnel's Mirrors

The set of sources, images, mirrors, and beams shown in Fig. 100 does not, I believe, require a great deal of explanation.

The function $F(x)$ is along the line AB, the angle between the normals of the two mirrors is equal to α, and the two images and the two mirrors are shown with superscripts corresponding to h' and h''. The two images are rotated with respect to each other by an angle 2α, and not by a fixed translation. The value of the angle 2α, the width AB of the slit, and the radius R of rotation must be such that the rotation be equivalent to a translation.

In general there are many advantages to choosing the width of the slit AB as narrow as possible: the slit may be considered as a line of points, and because the translation must be by at least the width of the slit, wide fringes are obtained near the mirrors. With a powerful light source and a wide slit,

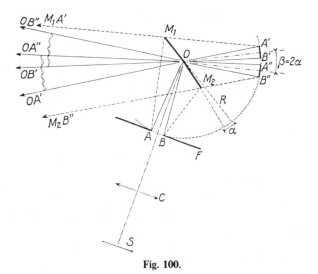

Fig. 100.

the fringes may be localized at fairly great distances. In 1923 Henri Buisson and I produced some at the end of a corridor 32 m long. In the same building of the Marseille Faculty of Science, P. Ruard has demonstrated them at 64 m.

With Fresnel's apparatus, it is not possible to adjust independently the width of $F(x)$ and the translation $h' - h''$. In particular, they may not be made equal to zero. We shall see in another example the distinctive features of interference by translation.

On a Remark by Wood

The remark in question was made by Wood about 1900 in response to the question: "Is it possible to observe interference beats between two spectral lines?"

Let a Michelson interferometer be illuminated with a plane wave of frequency ν'. Let the two reflected beams have the same intensity, equal to one, and the mirrors be adjusted for zero path difference $\delta = 0$. In the plane of the fixed mirror, which is taken as the observation plane, the time variation of the signal is

$$y' = \operatorname{cis} 2\pi\nu't. \tag{8}$$

If the path difference δ is increased, the amplitude y'' of the beam reflected in the moving mirror, with respect to the plane of the fixed mirror, is given by the classical equation

$$y'' = \operatorname{cis}\left(2\pi\nu't - \frac{\delta}{c}\right). \tag{9}$$

Let the moving mirror have a constant speed V, which increases the path difference δ when it is positive. The frequency of the reflected light is modified by the Doppler effect, and Eq. (9) becomes

$$\nu'' = \operatorname{cis} 2\pi\nu''t. \tag{10}$$

The two equations are strictly equivalent:

$$\delta = Vt \quad \text{and} \quad \nu' - \nu'' = \Delta\nu = \nu'\frac{V}{c}. \tag{11}$$

When Eq. (8) is added to Eq. (10), the result is an expression similar to Eq. (4), obtained for imaging:

$$y(t) = 2\operatorname{cis}\pi(\nu' + \nu'')t \cos\pi(\nu' - \nu'')t. \tag{12}$$

The frequency $\nu = (\nu' + \nu'')/2$ has replaced the linear variable x and the time t has replaced the angular variable u. Something has been gained: the range of t is unlimited, whereas that of u was limited by the circle of extension.

The interferometer could be illuminated otherwise, and the rings focalized. Equation (12) is valid only for the center of the rings. The beats ripple away from this center like wavelets when a stone is thrown into still water.

This remark of Wood's was reported to us by Ch. Fabry, who was there.

Spectral lines are not infinitely narrow: the frequency ν' is a point in the profile of the line $G(\nu)$, like the point P_0 in the slit AB of Fresnel's mirrors. If the line is narrow the translation $\Delta\nu$ is small. This difficulty is somewhat similar to that posed by the circle about which the images were rotated. In any case the result is well known: as the path difference δ increases, the visibility of the beats and of the fringes decreases. This visibility is the Fourier transform of the energy of $G(\nu)$. Indeed, the spectral line is an incoherent distribution within the limits of correlation functions, and interference fringes may only be observed for the frequency from one point of the line.

In interferometry the resolving power is usually defined by the separation of the two components of a symmetrical doublet like the one shown in Fig. 101a, for the yellow doublet of Sodium which often serves as an example. Let N be the mean frequency of the doublet, and ΔN be its width, defined by the first zero of visibility which, for the interference order n_0, is determined from

$$n_0 = \frac{N}{2\,\Delta N}. \tag{13}$$

The resolving power is equal to $2n_0$, and corresponds to a path difference δ smaller than one meter.

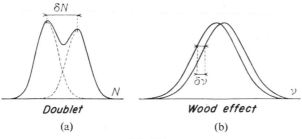

Doublet   Wood effect

(a)   (b)

Fig. 101.

It is a simple matter to impose on the delay δ a variation of one wavelength per second and to measure this delay as well as the propagation time over it with sufficient accuracy. If the wavelength is equal to 0.5 μm, the speed of the mirror is equal to 21.6 mm per day. Asparagus grows faster. The resolving power is

$$\frac{\nu}{\Delta\nu} = \frac{c}{u} = 1.2 \times 10^{19}. \tag{14}$$

For an incoherent doublet having the same width, the path difference corresponding to Eq. (13) would be $c/2 = 150,000$ km.

This is one of the paradoxes of resolving power in coherent imaging. There are others.

Symmetrical Pupils

Symmetrical pupils are those that have two apertures that are symmetrical with respect to the yz plane, illuminated by a system of waves with the same plane of symmetry. An example is the apparatus of Fizeau and Foucault, in which two slits parallel to the y-axis are placed between a collimator and a sight centered on the z-axis. Such pupils are relevant only in coherent illumination.

Let the two symmetrical pupils illustrated in complex space in Fig. 102 have the equivalent linear distributions $F(x)$ and $F(-x)$ along the x-axis. Let their transforms be broken up into their even and odd components:

$$*T[F(x)] = f_e(u) + f_o(u), \tag{15}$$

$$*T[F(-x)] = f_e(u) - f_0(u). \tag{16}$$

1. Let the two pupils be added:

$$*T[F(x) + F(-x)] = 2f_e(u). \tag{17}$$

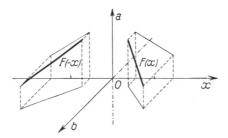

Fig. 102.

2. If the light incident on the pupil of Eq. (16) is retarded by half a period, the two linear distributions become $F(x)$ and $-F(-x)$, with

$$*T[-F(-x)] = -f_e(u) + f_o(u). \tag{18}$$

This is illustrated in Fig. (103). The sum of Eq. (15) and Eq. (18) is

$$*T[F(x) - F(-x)] = 2f_o(u). \tag{19}$$

3. If the two pupils are illuminated by plane waves, the two functions $F(x)$ and $F(-x)$ are real. Therefore $f_e(u)$ is real and $f_o(u)$ is imaginary. With a change of notation the Fourier transforms may be written

$$*T[F(x)] = f_e(u) + if_o(u), \tag{20}$$

$$*T[F(-x)] = f_e(u) - if_o(u). \tag{21}$$

The two Fourier transforms are complex conjugates. The real and imaginary parts of $F(x)$ are easy to extract from Eq. (17) and Eq. (18).

4. Let $F(x)$ again be a real function, with a phase advance of $\frac{1}{4}$ period. This is illustrated in Fig. 104a, and yields

$$*T[iF(x) + F(-x)] = \sqrt{2}[f_e(u) - f_o(u)] \operatorname{cis}\frac{\pi}{4}. \tag{22}$$

If the phase advance is given to the other aperture instead, then

$$*T[F(x) + iF(-x)] = \sqrt{2}[f_e(u) + f_o(u)] \operatorname{cis}\frac{\pi}{4}. \tag{23}$$

The sum and the difference of the real and imaginary parts of $f(u)$ may thus be determined optically.

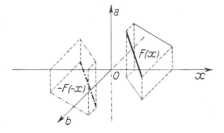

Fig. 103.

A Note on Symmetrical Pupils

Only symmetrical pupils without superposition are considered here. Let
$F(x)$ have the range $x > 0$ and $F(-x)$ have the range $x < 0$. In none of
the combinations considered, in the Fourier transform of the sum or of the
difference, have periodic zeros or cosine terms appeared, as they did in the
case of pupils replicated by a translation.

The set of the two pupils $F(x)$ and $F(-x)$ is the spectrum of their
common $f(u)$, x being considered as a frequency. This spectrum is even, and
its frequencies are both positive and negative. It is a simple matter to
conserve only the positive frequencies for its representation. The function
$F(-x)$ is simply omitted, and the ordinates of $F(x)$ are multiplied by two,
as shown in Fig. 104b.

Let $x'x''$ be the range of $F(x)$, and x_0 be the centroid of the distribution
$F(x)$ on the x-axis. It is not difficult to find a coefficient A so that

$$2F(x) = A\delta(x_0) \otimes F(x - x_0). \tag{24}$$

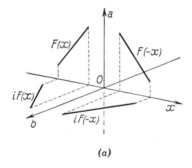

(a)

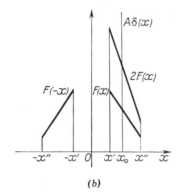

(b)   Fig. 104.

The Fourier transform of $F(x - x_0)$ is equal to $f'(u)$; that of $A\delta(x_0)$ is, in the representation of exclusively positive coordinates, equal to $A \cos 2\pi ux$. The Fourier transform may therefore be written

$$f(u) = f'(u) \cdot A \cos 2\pi ux_0. \tag{25}$$

The periodic fringes are modulated by the function $f'(u)$. The narrower the range $x'x''$, the slower the modulation.

Experiments illustrating this section will not be presented here.

STIGMATIC PUPILS

Stigmatic pupils are those that modify only the amplitude of the wave when they interrupt a train of plane waves. We consider here only the simplest cases, where a very thin screen with apertures is placed perpendicular to the direction of propagation of the waves. This may always be reduced to the case where the phase is equal to zero, and the complex amplitude distribution $F_0(x, y)$ in the plane of the screen is real. Its value is constant inside the apertures, and if it is taken equal to one, it becomes one of the range functions discussed in Chapter 5. This is illustrated in Fig. 105.

It is useful to study the Fourier transform $f(u, v)$ by means of sections with u or v = constant. For example take the section $v = 0$:

$$f(u, v) = \int\int_{-\infty}^{+\infty} F_0(x, y)\exp(-i2\pi ux) \, dx \, dy, \tag{26}$$

$$f(u, 0) = \int_{-\infty}^{+\infty} \exp(-i2\pi ux) \, dx \int_{-\infty}^{+\infty} F_0(x, y) \, dy. \tag{27}$$

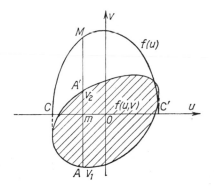

Fig. 105.

If we let

$$\int_{-\infty}^{+\infty} F_0(x, y)\, dy = F(x),\tag{28}$$

the expressions may be simplified:

$$f(u) = *T[F(x)].\tag{29}$$

The function $F(x)$ is the linear distribution equivalent to $F(x, y)$ for $v = 0$. It is a real and positive distribution. In Fig. 105, its value at the point m is seen to be a constant multiplied by the diameter AA'. Its range is the projection on the x-axis of the contour of the range of $F(x, y)$. We shall call the distance CC' along the x-axis the *width* of the aperture, although such a definition is unsatisfactory. It remains the same for all the equivalent linear distributions, when v is not equal to zero.

The real part of $f(u)$ is the spectrum of the real part of $F(x)$, and is an even function; its imaginary part is the spectrum of the odd part, and is odd. Because the function $F(x)$ is always positive, it may not be an odd function; $f(u)$ therefore always has a real part. For $f(u)$ to be real, $F(x)$ must be symmetrical with respect to Oy, that is to say, the diameters of $F(x, y)$ parallel to Oy must be equal to the abscissas $-x$ and $+x$. An example is shown in Fig. 106.

For $f(u)$ to be real for any direction of the x-axis, and therefore for any direction of the u-axis, the contour of $F(x, y)$ must be symmetrical with respect to the origin O. An example is shown in Fig. 107.

The Slit Pupil

We have already met this pupil in its usual form of a constant linear function $F(x)$. In fact, even when it is assumed to be unlimited in the

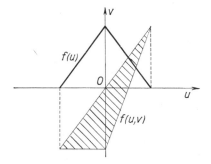

Fig. 106.

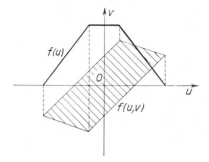

Fig. 107.

y-direction, it remains a two-dimensional diffraction function. Let $F_0(x, y)$ be a rectangular aperture with axes of symmetry Ox and Oy, having a complex amplitude density k. Let O be the center of symmetry. Let the range along x be from $-X$ to $+X$, and along y be from $-Y$ to $+Y$. The aperture is called a slit when Y is much greater than X.

The equivalent linear distributions are two rectangle functions. Let

$$\int_{-X}^{+X} F_0(x, y)\, dx = 2kX = A, \tag{30}$$

$$\int_{-Y}^{+Y} F_0(x, y)\, dy = 2kY = B. \tag{31}$$

The equivalent linear distributions are

$$F(x) = \begin{cases} B & \text{for} \quad -X < x < +X, \\ 0 & \text{elsewhere,} \end{cases} \tag{32}$$

$$F(y) = \begin{cases} A & \text{for} \quad -Y < y < +Y, \\ 0 & \text{elsewhere.} \end{cases} \tag{33}$$

Therefore

$$f(u) = 2BX \frac{\sin 2\pi Xu}{2\pi Xu}, \tag{34}$$

$$f(v) = 2AY \frac{\sin 2\pi Yv}{2\pi Yv}. \tag{35}$$

Taking note that

$$2BX = 2AY = 4kXY = kS, \tag{36}$$

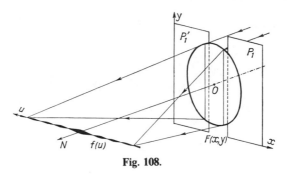

Fig. 108.

where S is the area of the slit, the Fourier transform of F_0 is

$$f_0(u, v) = kS \frac{\sin 2\pi Xu}{2\pi Xu} \frac{\sin 2\pi Yv}{2\pi Yv}. \tag{37}$$

The period of the zeros of $f(u)$ is equal to $1/X$, and the period of the zeros of $f(v)$ is equal to $1/Y$: the latter is the smaller by definition, and the diffraction appears to be linear. This is illustrated in fig. 108.

Sometimes the illumination of the slit is coherent along x and incoherent along y, for instance when the illumination comes from a circular symmetrical collimator illuminated by an incoherent narrow slit parallel to the y-axis. This source is almost a point source in the x-direction, and the light is therefore coherent along x, but it remains incoherent in the y-direction, or at the most partially coherent. The only zeros that remain are those along x. As shown in Fig. 109, all the coherent functions are parallel to x and to u.

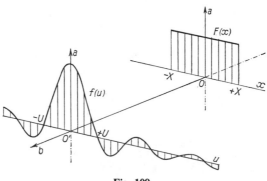

Fig. 109.

TWO-DIMENSIONAL PUPILS

We shall examine in detail only limited stigmatic pupils that have a nonzero complex amplitude transmittance over a finite range [that is, range functions $\varphi(u, v)$]; their amplitude spectra, the internal correlation functions $\Phi(x, y)$, are diffraction patterns, and we are concerned with the quadratic spectra of the squares of the moduli $|\Phi(x, y)|^2$:

$$\Delta(u, v) = \varphi(u, v) \otimes \varphi^*(-u, -v). \tag{38}$$

In Fig. 110 the plateau $\varphi(u, v)$ is hatched; $\Delta(u, v)$ is an image, and we have seen that there are two procedures for the image integration. They are considerably simplified in this case, because $\varphi(u, v)$ and $\varphi^*(-u, -v)$ are both real:

$$\Delta(u, v) = \varphi(u, v) \otimes \varphi(-u, -v). \tag{39}$$

First Procedure

The value at the point Q of the quadratic spectrum of the function of Fig. 110 is found by scanning over the whole range φ with the origin of the function $\varphi'(-u', -v') = \varphi(-u, -v)$ with axes Pu', Pv'. Two identical elements of integration ds' and ds are put at Q and P. The integration is then carried out over the range of φ for all the points Q inside the range.

Of course, the external contour of the range is the envelope of the contours of φ' when its origin P follows the external contour of the fixed range function φ.

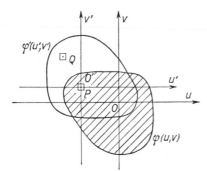

Fig. 110.

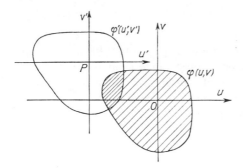

Fig. 111.

Second Procedure

For this procedure, illustrated in Fig. 111, a function $\varphi'(u, v)$ identical to $\varphi(u, v)$ is represented on axes u', v' parallel to u, v. Because the amplitude is equal to one in both the ranges φ and φ', the value of $\Delta(u, v)$ at the point P is equal to the area common to φ and φ', shown hatched in Fig. 111.

The contour of the quadratic spectrum is even easier to draw in this case: If the function φ' is moved about φ while the two ranges are kept in contact without allowing any overlap, the point P will draw the contour of the function $\Delta(u, v)$. The two ranges might have holes and concavities, but the functions in Fig. 111 and Fig. 112 do not have any.

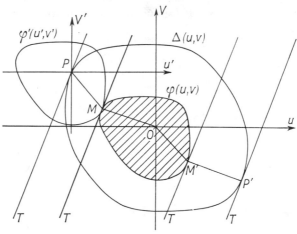

Fig. 112.

Let M be a point of contact between φ and φ'. If at least one of the two curves has a tangent at this point, the tangent to the contour of $\Delta(u, v)$ may be drawn without ambiguity. This construction finds its justification in theorems from kinematics. Let a tangent to φ at M' be drawn parallel to the tangent at M. It is obvious that the two vectors PM and OM are equal. Let $M'P'$ be drawn equal and parallel to MO; the point P' is also on the contour of Δ, and the tangent there is parallel to the three others. The contour of Δ is therefore an even function, and it does not require a great deal of thought to see that the function $\Delta(u, v)$ is also even. Of course it is also real.

If the integration of $\Delta(u, v)$ is carried out along the normal to the common tangent MT, the slope of the tangent plane to $\Delta(u, v)$ at M may be calculated. These questions have been studied in detail in an article published in 1945 in the *Revue d'Optique* by the author and Guy Lansraux, and could not be included in the first edition of this book.

The angle between the tangent plane and the uv plane depends on the curvature of the two contours at the point of contact. We shall only develop two examples in order to illustrate the main cases.

Quadratic Spectrum of the Square Pupil

The amplitude spectrum of a square pupil with side $2U$ is illustrated in Fig. 113. The contour of the quadratic spectrum is a square with the same center

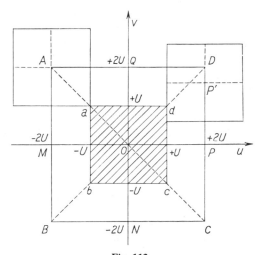

Fig. 113.

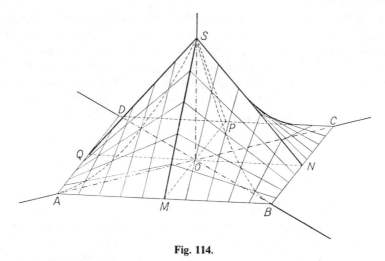

Fig. 114.

O and twice as big, having sides equal to $4U$. Besides the two squares φ (shown hatched) and Δ, two positions of the mobile square are shown, one at the corner A and the other along the side at an arbitrary point P'. The result of the integration is shown in Figs. 114 and 115, which show different sections of the function $\Delta(u, v)$.

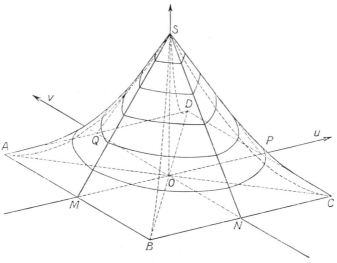

Fig. 115.

The vertex S is directly above the origin O; it is an absolute maximum. The network of straight lines of Fig. 114 correspond to squares displaced parallel to the u-axis. The lines MS, NS, PS, and QS are therefore straight lines. The displacement of the square along a diagonal obviously yields a parabola with the vertex at A, B, C, or D.

The exterior surface of $Z = \Delta(u, v)$ consists of four sections of identical hyperbolic paraboloids joined at the straight lines of the median planes. The sections of Fig. 115 are drawn parallel to the uv plane. Each is the vertex of an equilateral hyperbola which, when projected on the uv plane, would have the sides of the square $ABCD$ as asymptotes.

Quadratic Spectrum of a Circular Pupil

If the center of a circular aperture is at the origin of the axes u and v, the pupil has circular symmetry about the origin, and may have any radius smaller than $\mu = 1/\lambda$:

$$u^2 + v^2 \leqslant U^2, \qquad U \leqslant \mu = 1/\lambda. \tag{40}$$

Its diffraction pattern $F(x, y)$ also has circular symmetry; it is completely determined from the values along one radius, and is given by Eq. (6.64). The normalized diffraction pattern is shown in Fig. 116:

$$F(x,0) = \pi U^2 \frac{J_1(2\pi Ux)}{\pi Ux}. \tag{41}$$

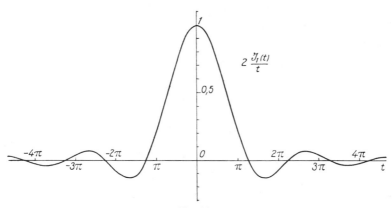

Fig. 116.

The quadratic spectrum also has circular symmetry. For the pupil having the maximum extension and an amplitude equal to one, the range function is

$$\varphi(u, v) = 1 \qquad \text{for} \quad u^2 + v^2 = \mu^2. \tag{42}$$

This is shown in Fig. 117 as the area with vertical hatching. The contour of the circle that encloses the quadratic spectrum $\Delta(u, v)$ is the locus of the centers P of the tangent circles with the same radius μ: it is therefore a circle with center O and radius equal to 2μ, or 2, if λ is taken as the unit of length.

At the point O' on the u-axis, $\Delta(u, 0)$ is the area common to the two circles with radii μ and centers at O and O'. This is the area of the two sectors with the common chord AA', shown with horizontal hatching. With

$$OH = \frac{u}{2} = \mu \cos \alpha, \tag{43}$$

it is easy to find an expression for $\Delta(u, 0)$ as a function of the parameter α:

area of sectors $OACA' = \pi \dfrac{2\alpha}{2\pi} = \alpha,$

area of triangle $OAA' = \dfrac{u}{2} \sin \alpha = \cos \alpha \sin \alpha = \dfrac{\sin 2\alpha}{2},$ $\qquad$ (44)

area of segment $ACA'H = \alpha - \dfrac{\sin 2\alpha}{2} = S,$

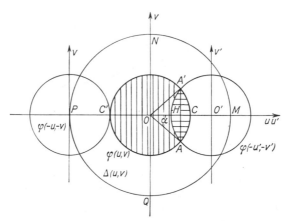

Fig. 117.

with

$$\Delta(u,0) = 2S = 2\alpha - \sin 2\alpha. \tag{45}$$

This function may easily be plotted point by point with great accuracy. In the logarithmic tables of François Callet, published in 1795 by Firmin Didot with his new stereotype process, a sine and cosine table graduated in steps of 0.1 grad is given with 15-digit accuracy, along with a conversion table from grads to radians, extended to 25 digits. Figure 118 reproduces the curve that we published with Guy Lansraux in the Revue d'optique in 1945, without numerical values.

The section of $\Delta(u, v)$ is therefore real and even; it has a cusp on the axis and is tangent to the uv plane along the circle with radius 2μ that limits its extension. The vertex corresponds to the exact superposition of the circles O and O'; its value is therefore equal to $\pi\mu^2$, that is to say π.

In order to study the tangent and the curvature, it is easier to move the origin of u to the point M with the positive direction towards O. Let the segment CH correspond to

$$u = 2h.$$

When compared with Eq. (45), this yields

$$\frac{d\Delta}{du} = \frac{dS}{dh} = S' = \frac{dS}{d\alpha}\frac{d\alpha}{dh}, \tag{46}$$

$$\frac{dS}{d\alpha} = 1 - \cos 2\alpha = 2\sin^2\alpha; \tag{47}$$

$$h = 1 - \cos\alpha, \quad \frac{dh}{d\alpha} = \sin\alpha, \tag{48}$$

$$S' = 2\sin\alpha. \tag{49}$$

1.   At the vertex having a value equal to π, the two tangents to the curve have slopes equal to $+2$ and -2. One is shown on Fig. 118, intersecting the axis at $u = \pi/2$. The value of $\sin\alpha$ is maximum at the vertex, and the curve $\Delta(u,0)$ remains very close to the tangent for most of its extent. For low frequencies, it may be replaced by the tangent with no serious loss in accuracy.

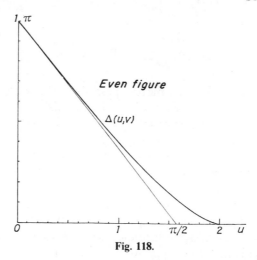

Fig. 118.

2. At the points M and P, $\sin \alpha$ is equal to zero, and the curve is tangent to the u-axis. At this point the curvature is given by the second derivative

$$\frac{dS'}{du} = \frac{2 \cos \alpha}{2 \sin \alpha} = \cot \alpha. \tag{50}$$

The curvature is infinite at points M and P: the slope is discontinuous there.

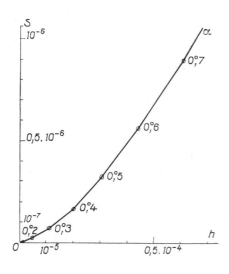

Fig. 119.

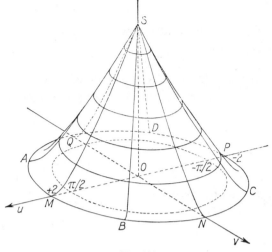

Fig. 120.

A discussion of a similar case may be found in the aforementioned article. In Fig. 119 is shown a strongly magnified and stretched version of the curve $S(h)$ for small values of the parameter α. With respect to Fig. 118, the magnifications are 4×10^4 for h and 2×10^6 for S.

The numerical data were used to draw Fig. 120, a perspective view of $\Delta(u, v)$ for a circular pupil.

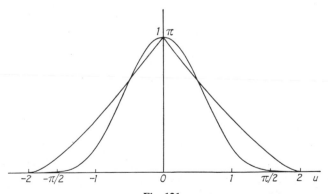

Fig. 121.

Apodizing Pupils

As an example, Fig. 121 shows the superposition of the quadratic spectra
for two pupils of revolution:

1. the circular aperture that we have just studied, and
2. the pupil with the same radius, but with a profile equal to $\cos \pi u$ inside
    the range $-1 < u < +1$.

It is curious that the apex of the quadratic spectrum depends directly on the
contour of the amplitude spectrum. The high-frequency rolloff favors the
transmission of low frequencies, that is, the concentration of the light beam.
This is particularly true for laser light.

9

Discrete Functions

INTRODUCTION

The title of this chapter may appear a bit extravagant, but I think that it emphasizes the main concepts of this chapter, which include the elementary theory of gratings. Of course the extension of the concept of the Laue sphere to periodical spectra of light distributions makes the circle of radius $\mu = 1/\lambda$, which is its extreme limit, appear in a new light. This circle is no longer an arbitrary section of an indefinite frequency plane u, v, but the equator of a singular Laue sphere. This view is clarified by the introduction of Lambert's law and the precise relation between the three transforms of the range functions in one, two, and three dimensions.

In the discrete case the Fourier transform of the spectrum $f(u, v)$ is a series and the previously continuous function $F(x, y)$ becomes a discrete ensemble of points, which we call the *discrete function* $F_0(x, y)$. The function $F(x, y)$ is thus stripped of the redundancies that made us call it a diffracted function when we discussed Dirichlet's theorem. The pupil function is allowed to be spatially limited, while satisfying the principle of conservation of energy and giving to $f(u, v)$ its true interpretation as a limited probability function whose spectrum is also limited. If the wave is considered not as physically real, but as a mechanical, mathematical, or analogical model, the use of a certain amount of mathematical license may allow the consideration of certain distributions that the diffracted function $F(x, y)$ does not allow. To emphasize the fact that the points of $F_0(x, y)$ are not physically real, we call them *information points* because they have no redundancy. The whole set F_0 makes up an information image.

As a matter of fact, these points are not resolved under the conditions under which they will be defined, but they are themselves too large and too far apart to resolve the structure of the photon. The smallest separable point in optical imaging is the central fringe of the range function of Laue's sphere of extension: a sphere with a diameter equal to λ. But in the focused beam of a pulsed laser, there are 1000 or even 10,000 photons in this sphere. It must be concluded that in the final analysis, linear optical imaging, interference, diffraction, and linear reflections are powerless to yield a complete understanding of what light is. In order to attain the deep structure, a higher resolving power is required, such as that of atoms and molecules, which emit and absorb photons and modify their polarization and its energy. But this would draw us into nonlinear optics, where the interaction of light with matter is the tool of the trade.

This chapter is almost exclusively dedicated to coherent images of one variable. The theory of discrete images developed here may not be applied to incoherent imagery without certain precautions and certain changes to be mentioned at the end. There must be some possible combinations between coherent and incoherent discrete images, but this would lead us far astray without adding anything to the possibilities of the Fourier transform.

We shall see in the next chapter some simple relations between distribution functions in one, two, and three variables. In optics that is really physical, continuous distributions are only (and barely) justified by special relativity and Lorentz's equations: the observer sees the photon as a discontinuity, and is himself seen as a discontinuity by the photon. Whenever matter intervenes, no surfaces are without thickness. In fact, Professor Inglestam has given a relationship between the uncertainties in a topography and the relief corresponding to an interference distribution.

THE CONVERSE OF DIRICHLET'S THEOREM

In Chapter 5, the application of Dirichlet's theorem to bandlimited functions led to two general expressions:

$$f\varphi(\cdot) = f, \qquad (5.50)$$

$$F \otimes \Phi(\cdot) = F. \qquad (5.51)$$

The function F is a distribution of one, two, or three variables; f is its spectrum or its Fourier transform, φ is the range of f, and $\varphi(\cdot)$ is its range function, whose value is equal to one over the range φ. $\Phi(\cdot)$ is the Fourier transform of $\varphi(\cdot)$. F is a diffracted function of wavelike origin, whereas f is a Euclidean function with corpuscular implications.

In classical physical optics, F is given and f is the result of the analysis. In particle optics it is natural to consider the photon flux f as primary, either given precisely or in terms of probability, and the problem is then to find a Euclidean F_0 that satisfies the relation

$$F = F_0 \otimes \Phi(\cdot). \qquad (1)$$

This may be solved easily.

Functions of One Variable

Whatever the number of variables, the problem remains the same. In fact the function f has only two independent variables, and we have discussed the experimental complications brought about by the three-dimensional functions that are met with in x-ray diffraction. For a function of one variable, the solution is much simpler. We shall therefore consider the simplest form of Eqs. (5.50) and (5.51):

$$f(u)\varphi(u) = f(u),\tag{2}$$

$$F(x) \otimes \Phi(x) = F(x),\tag{3}$$

where

$$\varphi(u) = \begin{cases} 1 & \text{for } u \in U, \\ 0 & \text{elsewhere.} \end{cases}\tag{4}$$

When $\varphi(u)$ is even,

$$\Phi(x) = U\frac{\sin \pi Ux}{\pi Ux}.\tag{5}$$

If $\varphi(u)$ is shifted by h along the u-axis, then $\Phi(x)$ is multiplied by $\exp(-2\pi ihx)$, a helical twist or phase shift that leaves the zeros and the modulus unchanged. The following calculations are for an even $\varphi(u)$.
In order to obtain $F_0(x)$ that satisfies

$$F_0(x) \otimes \Phi(x) = F(x),\tag{6}$$

it is sufficient to take for the function $F_0(x)$ the set of terms from the Fourier series that represents $f(u)$ replicated over its range.
Indeed, consider the series in the form of Eq. (2.29), but with reciprocal variables

$$f(u) = \sum_{n+-\infty}^{+\infty} Z_n \operatorname{cis}(2\pi nXu), \quad \text{where} \quad X = \frac{1}{U},\tag{7}$$

in which U is the range of $f(u)$, and

$$Z_n = X\int_U f(u)\operatorname{cis}(2\pi nXu)\, du.\tag{8}$$

Let us write the set $F_0(x)$ as

$$F_0(x) = \{Z_n\}_{n=-\infty}^{+\infty}. \tag{9}$$

The function $f(u)$ may also be represented as an isolated function over the whole range of the variable u, by a Fourier integral that is precisely equal to $F(x)$:

$$F(x) = \int_{-\infty}^{+\infty} f(u)\operatorname{cis}(2\pi xu)\, du, \tag{10}$$

$$F(x) = \int_U f(u)\operatorname{cis}(2\pi xu)\, du. \tag{11}$$

Comparing this with Eq. (8) yields

$$Z_n = XF(nX). \tag{12}$$

This is the same relation that was given in Chapter 1 when the relations between series and integral representations of the same solitary function were discussed.

The convolution of Eq. (6) is equal to

$$F_0(x) \otimes \Phi(x) = \sum_{n=-\infty}^{n=+\infty} Z_n U \frac{\sin \pi U(x - nX)}{\pi U(x - nx)} \tag{13}$$

$$= UX \sum_{n=-\infty}^{+\infty} F(nX) \frac{\sin \pi U(x - nX)}{\pi U(x - nX)} \tag{14}$$

$$= \sum_{n=-\infty}^{+\infty} F(nX) \frac{\sin \pi U(x - nX)}{\pi U(x - nX)}, \tag{15}$$

because

$$UX = 1. \tag{16}$$

The zeros of $\Phi(x)$ coincide with the terms Z_n, except for $n = 0$, when $\Phi(x) = U$, as shown in Fig. 122. The terms of order $k \neq n$ have no effect on the term of order n; the function that results from the convolution (6) therefore coincides with $F(x)$ at every point where there is a term of the series; it has the same spectrum, and it is therefore equal to $F(x)$ over its range.

In Fig. 122, $F(x)$ is a real amplitude distribution in the diffraction fringes whose shortest period is the same X that appears in $(\sin \pi Ux)/\pi Ux$. It is the dashed curve in the figure. The ordinates of $F(nX)$ are also shown in the figure.

The reciprocal relations between F_0 and f are

$$F_0(x) = \{Z_n\}_{N'}^{N''} = X\left\{\int_U f(u)\text{cis}(2\pi nXu)\,du\right\}_{N'}^{N''}, \qquad (17)$$

where n is an integer, and

$$f(u) = \sum_{N'}^{N''} Z_n \text{cis}(-2\pi nXu), \qquad \text{with} \quad UX = 1. \qquad (18)$$

Of course these two equations satisfy Plancherel's theorem:

$$\sum_{N'}^{N''} |Z_n|^2 = \int_U |f(u)|^2\,du. \qquad (19)$$

RESOLUTION OF A SET OF POINTS ALONG A LINE

The separation of the images of two points having equal intensity is a problem that has been much considered in instrumental optics. Lord Rayleigh's solution, which imitated the resolving power of the retina, has never been considered satisfactory. The image of two points that are not exactly superimposed is never identical to the image of an isolated point.

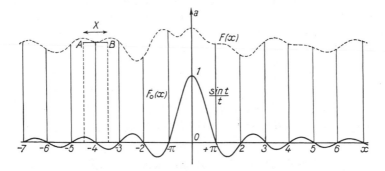

Fig. 122.

However, in incoherent light, the smaller the distance between the points, the more difficult it is to determine it. With coherent light, the separation is always easy if a phase difference may be introduced between the two points. If the phase difference is equal to π, when the two points coincide the two images will cancel each other out by interference, and it will be easy to determine the distance between the points.

The Fourier transform, the Foucault gratings that will be described in the next chapter, and the grating theory that we shall develop will lead us to consider the problem not for two isolated points in a given area, but for a finite set of points over a finite domain, which may be either isolated or replicated periodically.

Bounded Sets

N points are randomly without overlap in a region p. The region p and its interior can be represented by a function or with a periodicity analyzed into a periodic distribution. The object and its image, reduced to the same variable x, are represented by a Fourier series whose fundamental frequency is $1/p$:

$$U_0 = \frac{1}{p}. \tag{20}$$

The series representing the object is infinite and nonconvergent. If I_k is the fixed magnitude of a point, then

$$F_0(x) = \sum_{m=-\infty}^{+\infty} \sum_{k=1}^{N} I_k \operatorname{cis} 2\pi m U_0 (x - x_k). \tag{21}$$

Incoherent Illumination

I_k, which is the illumination energy of a point of order k, is real and positive. Its incoherent image $F(x)$ is a distribution modulated by the transmission factor $d(u)$ of the optical system of form:

$$F(x) = \sum_{m=-M}^{+M} \sum_{k=1}^{N} d(m U_0) \cdot I_k \operatorname{cis} 2\pi m U_0 (x - x_k), \tag{22}$$

where, the range of $d(u)$ being finite, $-M < m < +M$.

Experiment yields this series in the form

$$F(x) = \sum_{m=-M}^{+M} Z(m U_0) \operatorname{cis} 2\pi m U_0 X. \tag{23}$$

The unknown parameters of the image are determined from the equations

$$Z(m) = \sum_{k=1}^{N} d(mU_0) \cdot I_k \operatorname{cis}(-2\pi m U_0 x_k), \quad -M < m < +M. \quad (24)$$

The functions $F_0(x)$, $F(x)$, and the diffration pattern $e(x)$, which is the Fourier transform of $d(u)$, are all real functions. Their spectra may have an even real part and an imaginary odd part. Among the equations (24) there are only $M + 1$ distinct equations, including the zero-frequency terms, which are real and which allow the determination of the units to be used to measure the energies (or intensities) and the moduli:

$$Z_0 = \sum I_k. \quad (25)$$

On taking into account the real and imaginary parts of the M remaining equations, there remain $2M$ equations to determine the unknowns I_k and x_k, that is, $2N$ unknowns. The image may be solved for if

$$N \leqslant M; \quad (26)$$

that is, if

$$p \geqslant N\varepsilon, \quad (27)$$

where ε is the smallest period transmitted in the range of $d(u)$. The value of ε is the classical limit of resolution. If the range $d(u)$ is given its maximum value 2μ, the limit is equal to $\lambda/2$, but in incoherent imagery, it is the average distance between the points that must not be smaller than the resolving power.

Coherent Illumination

Consider the same number of points N distributed at the locations x_k within the same range p, with complex amplitudes A_k having the moduli $|A_k| = + \sqrt{I_k}$ and phases ψ_k. Then $F_0(x)$ is a complex function, as is its image $f(x)$. If we want to follow a route parallel to that of the previous section, we must take as the transmission function $h(u)$, the function from which $d(u)$ was derived by autocorrelation. Figure 97b at the end of Chapter 7 reminds us that if the range of $h(u)$ is U, the range of $d(u)$ is equal to $2U$. But in general no symmetry may cancel out any part of this range. The range of frequencies that may yield equations for resolving the points is therefore the same for the coherent and incoherent cases. There are therefore $2M + 1$

equations given by the real and imaginary parts of $h(u)$, equations similar to (24).

But the coherent points need half again as much information as the same number of incoherent points: we must determine the abscissas x_k and the real and imaginary parts A'_k and A''_k, that is, three values per point instead of two. The system of N coherent points that occupy the same range p and whose images are formed by the same optical instrument is therefore not resolvable under conditions (26) and (27).

If we have previously found that the minimum period was the same X for both coherent or incoherent illumination, it is because the question of the abscissas did not arise for a set of periodic points starting at an arbitrary origin. In sum, it is possible to calculate the complex amplitudes of M points with a period $X = 1/U$.

Information Points

This is the name that we give to the periodic points of the function F_0. Whether they belong to a coherent amplitude distribution or to an incoherent distribution of energy, they are at the limit of what we now call the resolving power, that is, the smallest frequency corresponding to the spectral range U. When U has its maximum value $2/\lambda$, the period X of the information points is equal to $\lambda/2$. To maintain the possibility of nonmaximum values of U, which is the most frequent case, we abandon the classical notation with ε; X is not a vanishing quantity.

In a given pupil, the information points are at the same locations whether the illumination is coherent or incoherent. One should consider the possibility that they are at the same places in partially coherent illumination.

General Remarks

1. Equation (7) yields $f(u)$ from the series Z_n; let us recast it by replacing the Z_n with their values (12), and write underneath it the integral giving the same $f(u)$ as a function of $F(x)$:

$$f(u) = \sum_{n=-\infty}^{+\infty} XF(nX)\mathrm{cis}(-2\pi nXu), \qquad (28)$$

$$f(u) = \int_{-\infty}^{+\infty} F(x)\mathrm{cis}(-2\pi xu)\, dx. \qquad (29)$$

The expression (28) is an approximation to the integral (29). Usually the approximation differs from the integral by an error term. Here the error term is *equal to zero* for the whole range of integration. It is possible that a partial sum between x_0 and x_1 may differ from the integral over the same limits, even if $x_0 = n_0 X$ and $x_1 = n_1 X$.

2. Change the notation on the right side of Eq. (11), and take the first derivative of $F(x)$:

$$F(x) = \int_U f(u)\exp(i2\pi ux)\, dx, \tag{30}$$

$$\frac{dF(x)}{dx} = i2\pi \int_U uf(u)\exp(i2\pi ux)\, du. \tag{31}$$

Now $dF(x)/dx$ has the Fourier transform $2\pi if(u)$, which is also limited to the range U, and which may therefore be replaced by a discrete function whose information points coincide with those of $F_0(x)$. This new discrete function $[dF(x)/dx]_0$ *may therefore be called the first derivative of* $F(x)$. We call this function $F_0'(x)$. The same may be done for higher-order derivatives, as long as the traditional derivative of $F(x)$ exists. These derivatives may be called $[d_k F(x)/dx^k]_0$. The methods of calculations applied to $F(x)$ may easily be applied to $d_k F(x)/dx^k$.

3. Let us continue with coherent imaging, which will bring about no difficulties. Equation (13) and (15), whose convolutions represent the function $F(x)$ (the fundamental function for mathematical and physical definitions), offer two processes of integration.

Equation (13) is derived directly from Dirichlet's theorem. The Z_n of $F_0(x)$ are given by Eq. (12), which gives to the points $F(nX)$ a range X. If x_0 is the abscissa of the point, the range may be limited to $x_0 - X/2$ and $x_0 + X/2$. In Fig. 122, the range of the point at $x_0 = -4X$ is shown. The rectangle bounded by the segment AB integrates the part of $F(x)$ attributed to Z_{-4}. It should be noted that the modulus and the phase in this range are uniformly those of $F(nX)$: the range is unresolved. The dissipation function and internal correlation function of Eq. (13) is $\Phi(x)$, the Fourier transform of the range function of $f(u)$.

Equation (15) is the simplest form of the convolution. $F(x)$ is represented by the point nX, the range function having been normalized, and the internal correlation function is reduced to $(\sin \pi U_x)/\pi Ux$. This is the loss function that is represented by Fig. 122, when the construction about the point $-4X$ is removed.

4. The energy distribution corresponding to the amplitude distribution is proportional to

$$D(x) = |F(x)|^2. \tag{32}$$

The information points I_n of $D_0(x)$ coincide with those of $F_0(x)$, and we must distribute $|F(x)|^2$ among the former as we have distributed $F(x)$ among the latter:

$$I_n = XD(nX) = X|F(nX)|^2. \tag{33}$$

Now the normalized amplitude element $\Phi(x)$ must be replaced by a normalized energy element derived from $|\Phi(x)|^2$. The result is

$$\int_{-\infty}^{+\infty} \frac{\sin t}{t}\, dt = \int_{-\infty}^{+\infty} \left(\frac{\sin t}{t}\right)^2 dt = \pi. \tag{34}$$

Replacing t by πUx yields

$$U\frac{\sin \pi Ux}{\pi Ux} = U\left(\frac{\sin \pi Ux}{\pi Ux}\right)^2 = 1. \tag{35}$$

Let

$$\Phi'(x) = U\left(\frac{\sin \pi Ux}{\pi Ux}\right)^2. \tag{36}$$

Three expressions similar to Eqs. (13), (14), and (15) may be written for $D(x)$:

$$D(x) = D_0(x) \otimes \Phi'(x) = \sum_{n=-\infty}^{+\infty} I_n U\left(\frac{\sin \pi U(x - nX)}{\pi U(x - nX)}\right)^2 \tag{37}$$

$$= UX \sum_{n=-\infty}^{+\infty} |F(nX)|^2 \left(\frac{\sin \pi U(x - nX)}{\pi U(x - nX)}\right)^2 \tag{38}$$

$$= \sum_{n=-\infty}^{+\infty} |F(nX)|^2 \left(\frac{\sin \pi U(x - nX)}{\pi U(x - nX)}\right)^2. \tag{39}$$

The simplicity and the similarity between the two expressions (15) and (39) is quite remarkable. In Fig. 123 are shown all the spectra of interest, in imitation of Fig. 97 of Chapter 7.

The step AB with unit height and with length equal to U is the spectrum of $\Phi(x)$. The triangle with the base $-U, +U$ with its apex at $+U$ is the spectrum of $|\Phi(x)|^2$. The triangle with the same base whose apex is at 1 is the spectrum of $\Phi'(x)$; it has the same area as the step AB. The step $A'B'$,

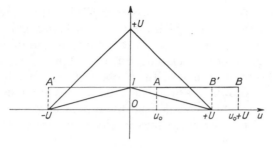

Fig. 123.

which Dirichlet's theorem would suggest, has no immediate interest. However, it leaves intact all the spectra of $D(x)$, whether they are integrals or series.

In Fig. 124 are shown superimposed the two integrated curves of Eq. (34), which (after a simple scale change) are the two functions of Eq. (35). One represents the unit of complex amplitude, and the other the unit of energy. The energy is much more concentrated about the origin.

LIMITED DISCRETE PUPILS

We shall continue to treat the information points as being at the limit of resolution, with the period X larger than $\lambda/2$, its value being determined by the transmission of the optical instrument under consideration. It is obvious that functions F_0 having one, two, or three dimensions may exist inside an infinite domain Δ. Three resolvable coherent points yield at infinity a diffraction pattern with interference fringes. If the number of points is increased until the resolution limit is reached, then for a given extent of the spectrum, neither the period X nor the internal correlation function $\Phi(\cdot)$ will change.

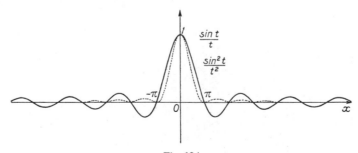

Fig. 124.

However, the use of discrete coherent pupils is not without its problems, associated with the phase, which must be distributed among the photons according to some law that is not yet clear. Incoherent points also depend on the wavelength λ of the radiation, but they allow a random distribution, like the photons of linear optics. The photoelectronic chambers of Professor Lallemand, well known to astronomers, allow very significant experiments in this respect. A light image is incident in a vacuum on a photosensitive plate that transforms the image into an electronic image. The transmission between the two images has a very high resolving power. The beam of electrons is accelerated and concentrated into a stigmatic image on a photographic plate. The object and the electronic image are incoherent, and would be so even if the light image were coherent. In the thirties, a similar phenomenon was used by spectroscopists in the far ultraviolet. Gelatin silver bromide plates were not very sensitive in this range. The emulsion was covered by a thin layer of paraffin oil, whose fluorescence displaced the quanta towards frequencies at which the plate was more sensitive. Here again a coherent image became incoherent while remaining stigmatic.

Experiments of this type have maintained our conviction that the definition of coherence given by Young and Fresnel remains fundamental: it is the only one that may be used in particle optics.

Two-Dimensional Pupils

A one-dimensional function may be considered to be derived from a two-dimensional function, which is generated by translating the x-axis along y. There is no variation along y, so the function is essentially a function of x only: $F(x)$. In Fig. 125, we have displayed a real distribution of information points in a thin slit having a width of $41\lambda/2$; such a slit may be generated by evaporating metal onto a transparent substrate. The hemisphere of

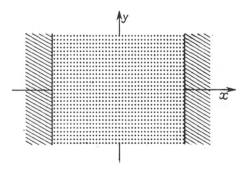

Fig. 125.

extension is subjected to diffraction, and along a long line of points in the y-direction, the internal correlation functions (or impulse responses) are superimposed in a continuous unresolvable series. The slit is assumed to be narrow enough so that the diffraction from one edge is influenced by the diffraction by the other edge.

The advantage of the slit is that the effects of diffraction at the edges are well removed from the region of interest. They may be attenuated by reducing the illumination near the edges.

Circular Pupils

The use of limited two-dimensional pupils with arbitrary shapes is not without its difficulties. Most of the problems may be examined by considering the case of a pupil limited by a circular aperture. We assume that the spectrum remains confined within the circle of extension with radius $\mu = 1/\lambda$.

In Fig. 126a, such a pupil is shown with Cartesian coordinates Ox and Oy. Because the spectrum of $F(x, y)$ is limited, $F(x, y)$ has an unlimited extent, but $F_0(x, y)$ is a set of points disposed in a regular array in two dimensions limited to the circular pupil with center O and radius R. The points Z are distributed in arrays parallel to the coordinate axes. In imitation of Eqs. (9), (12), (13), and (15), we write

$$F_0(x, y) = \{Z_{nm}\} \qquad \text{with} \quad n^2 X^2 + m^2 Y^2 \leqslant R^2, \tag{40}$$

$$Z_{nm} = XY \cdot F(nX, mY) \qquad \text{with} \quad UX = VY = 1, \tag{41}$$

$$F_0(xy) \otimes \Phi(xy) = \sum\sum Z_{nm} UV \frac{\sin \pi U(x - nX)}{\pi U(x - nX)} \frac{\sin \pi V(y - mY)}{\pi V(y - mY)} \tag{42}$$

$$= \sum\sum F(nX, mY) \frac{\sin \pi U(x - nX)}{\pi U(x - nX)} \frac{\sin \pi V(y - mY)}{\pi V(y - mY)}. \tag{43}$$

U and V may be determined by experimental conditions to be below their maximum values 2μ. When they are unspecified, we shall use this value. X and Y have the same value $\lambda/2$, and the domains of the information points are squares defined by

$$n^2 + m^2 \leqslant 4 \frac{R^2}{\lambda^2}. \tag{44}$$

If the limits of its domain are drawn about each point, the circumference is replaced by a broken line unresolved in x and y.

The initial circular symmetry has been replaced by a rectangular symmetry. If we went into three dimensions, the spherical symmetry so far presented by the properties of light corresponding to an isotropic homogeneous space would be replaced by a rectangular symmetry. For a circular pupil the normal correlation function is the Airy disk, or if a third dimension is introduced, the spherical function $(\sin 2\pi ur)/2\pi ur$, where r is measured in units of λ. The photon (or the aspect of it that appears in linear optics) seems to have this circular symmetry.

A closer approximation to this symmetry is the hexagonal array shown in Fig. 126c. There exists a Fourier transform for oblique axes, and the figure shows one of the possible sets of axes. (The reader is referred to texts on crystallography.) The domain of each point is a hexagon, and all the points are at a distance $\lambda/2$ from their six neighbors. In three-dimensional optics, the corresponding figure would be an icosahedron, which corresponds to the most stable configuration. The circle of Fig. 126a contains 301 points, whereas the circle of Fig. 126c contains 265. But the $\lambda/2$ of the hexagons is greater than that of the squares, and for the same wavelength, the hexagon would contain 331 points. That the hexagonal system is the most stable may also be demonstrated by piling up balls.

In Fig. 126b the 301 points of Fig. 126a are distributed at random. This was done by sprinkling a fine spherical powder called floramine on a sticky sheet of paper. This is a possible distribution of efficient photons on the screen of one of Professor Lallemand's photoelectronic chambers. This distribution may not be used to represent an unresolved complex amplitude distribution, but it is acceptable for a uniform probability distribution of energy.

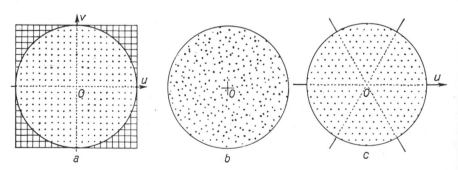

Fig. 126.

LIMITED PERIODIC FUNCTIONS—GRATINGS

Ever since Fraunhofer, gratings have been much studied. We shall consider them only within the mathematical context that we have studied so far, that is to say, from a very simplistic point of view. The fundamental theorem was introduced in Chapter 1 under the title "Isolated functions and associated periodic functions." The integrals and series seen there were unlimited. By the use of the δ-function and of Dirichlet's theorem and its converse, which we are now acquainted with, we can understand more complex aspects that are closer to reality.

Series of δ-functions

Let the initial function be a normalized gaussian $F(x)$; its spectrum is another gaussian function:

$$F(x) = \frac{\sqrt{\pi}}{a} \exp\left(-\frac{\pi}{a}x^2\right), \tag{45}$$

$$f(u) = \exp(-\pi a^2 x^2). \tag{46}$$

Because $F(x)$ is normalized, the value of $f(u)$ at the origin is fixed: $f(0) = 1$. This is shown in Fig. 127.

The associated periodic function $G(x)$, with period X, is easy to construct: it is a function $\theta_3(x)$ (Fig. 43b), but it is not shown in Fig. 128, where only the maxima and their ordinates at periodic intervals are shown. The Fourier transform of $F(x)$, one period of $G(x)$, becomes a set of Z_m

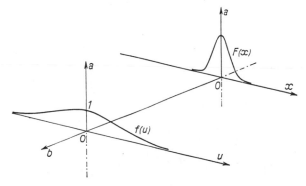

Fig. 127.

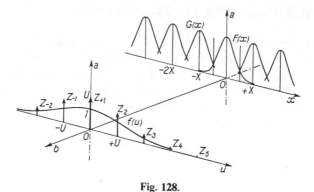

Fig. 128.

terms:

$$\{Z_m\}_{m=-\infty}^{m=+\infty} = \{Uf(mU)\}_{-\infty}^{+\infty}, \qquad \text{with} \quad UX = 1. \tag{47}$$

If a goes to zero, then $F(x)$ tends to a delta function $\delta(x)$ and $f(u)$ tends to a function $H_0(0)$. At the same time $G(x)$ tends to a periodic function consisting of a series of functions $\delta(x - nX)$, while $\{Z_m\}$ tends to a series of points with ordinates mU, period U, and amplitudes U. This Fourier series of a periodic series of delta functions is written

$$T[\delta(x)] = f_0(u) = \{U_m\}_{-\infty}^{+\infty}. \tag{48}$$

This is shown in Fig. 129.

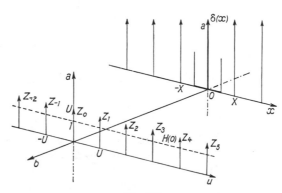

Fig. 129.

But note that the points Z_m are not integrable points but finite densities. Let the center of the period X of $\delta(x)$ be at $x = 0$:

$$\delta(x) = \sum_{m=-\infty}^{+\infty} Z_m. \tag{49}$$

This equation is parallel to

$$F(0) = \int_{-\infty}^{+\infty} f(u)\, du. \tag{50}$$

Mathematical Construction of a Finite Grating

This will be accomplished by a series of convolutions. To avoid confusion the following notation will be followed:

$G(x)$ is the periodic function corresponding to the grating (at the start it will be unlimited, but later on limited);

$F(x)$ represents the content of one period X of the grating, centered upon the origin of x: $-X/2 < x < +X/2$ (the origin can be at the beginning of a period: $0 < x < X$ without changing the conclusions);

$f(u)$ is the integral Fourier transform of $F(x)$, and $f_0(u)$ is its series transform;

$g(u)$ will be the Fourier transform of the limited grating, when we finally get it.

Starting Point

We begin with the unlimited array of functions of Fig. 129. The series $G(x)$ is the set

$$G(x) = \{\delta(x - nX)\}_{n=-\infty}^{n=+\infty}, \qquad n \text{ an integer.} \tag{51}$$

The content of the period corresponding to $n = 0$, centered on the origin and containing $F(x)$, is $\delta(x)$; its integral transform $f(u)$ is a function $H(0)$, and its series transform $f_0(u)$ is a periodic set of vectors U with period U, denoted by U_m, with $UX = 1$:

$$f_0(u) = Uf(mU) = \{U_m\}_{m=-\infty}^{m+=\infty} \qquad m \text{ an integer.} \tag{52}$$

First Operation

A new $G(x)$ is formed by convolving with $\delta(x)$ a function $F(x)$ that is integrable and generally continuous. This *isolated function* has a Fourier transform $f(u)$:

$$T[F(x)] = f(u). \tag{53}$$

The result of the convolution is that all the terms U_m of Eq. (52) are multiplied by the Fourier transform of Eq. (53). The new $f_0(u)$ is

$$f_0(u) = \{Uf(mU)\}_{m=-\infty}^{m=+\infty} = \{Z_m\}_{m=-\infty}^{m=+\infty}. \tag{54}$$

The Z_m are the information points of $f(u)$ reduced to the interval X. There are two cases that deserve discussion:

1. The elementary function $F(x)$ is entirely contained inside one period X. The convolution therefore leaves it intact:

$$G(x) = \{\delta(x - nX)\}_{n=-\infty}^{n=+\infty} \otimes F(x). \tag{55}$$

A representation of this convolution is given in Fig. 1 of Chapter 1.

2. The elementary function $F'(x)$ is wider than one period X. This is the case in Fig. 12 of Chapter 1. The mechanism of convolution is explained in Fig. 13. There are two methods of calculation. First calculate $F(x)$:

$$F(x) = \sum_{n=-\infty}^{n=+\infty} F'(x - nX) \qquad \text{for} \quad -X/2 < x < +X/2. \tag{56}$$

This yields the convolution (55) directly. But Fig. 13 clearly shows that $F'(x)$ may be introduced directly into the convolution:

$$G(x) = \{\delta(x - nX)\}_{n=-\infty}^{n=+\infty} \otimes F'(x). \tag{57}$$

The treatments are exact, because $G(x)$ is still an infinite series. $F(x)$ may also be unlimited.

Second Operation

Let us now limit the array to N periods. We shall need a simple theorem illustrated by Fig. 130, which shows a linear grating consisting of modulated rectangular apertures along with its spectrum. The latter has all the desirable qualities: it is unlimited, and its complex amplitudes follow a left-handed helix (a linear phase shift) in complex space. It represents the period between $-X/2$ and $+X/2$.

By means of the translation nX, let this period and its content be brought into coincidence with the period of order n. The spectrum is affected by a helical twist or phase shift $\mathrm{cis}(2\pi nXu)$. The term Z_m of the spectrum of Fig. 130 is multiplied by the twist (phase) factor

$$\mathrm{cis}(2\pi nmXU) = \mathrm{cis}(2\pi nm) = 1, \qquad (58)$$

because $XU = 1$ for any system of units. All the Z_m have come back to their original places. *The graphical spectrum $\{Z_m\}$ of the period $F(x)$ is the same for any period of $G(x)$.*

The information points of the spectrum of the N periods of the limited G(x) are obtained by multiplying by N the terms Z_m of $f_0(u)$ and the vectors that represent them.

This statement identifies the series spectrum of $G(x)$ in its limited domain $L = NX$ with that of the N periods superimposed on $F(x)$. It will temporarily suffice to define the domains of the information points: the domain of the point Z_m of $f_0(u)$ is equal to U; the domain of the point NZ_m of the spectrum $g_0(u)$ is reduced to U/N.

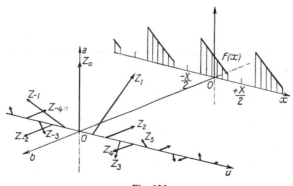

Fig. 130.

Let $\Sigma^N F_n(x)$ be the N aligned periods of the array. Let their range be L, the range function be $L(x)$, and its Fourier transform $l(x)$. Then

$$L = NX,$$

$$l(u) = NX\frac{\sin \pi NXu}{\pi NXu}.$$

(59)

If ΔU is the period of the zeros of $l(u)$, then

$$L \Delta U = 1.$$

(60)

But

$$NX = \frac{N}{U},$$

(61)

so that

$$\Delta U = \frac{U}{N}.$$

(62)

This is equal to the domain of the information points of $g(u)$, but they coincide with the zeros of this function except for the NZ_m. To obtain the light distribution in $g(u)$ from the information points, they must be convolved with the Fourier transform $l(u)$ of the range of $G(x)$:

$$g(u) = Nf_0(u) \otimes l(u) = \sum_{m=-\infty}^{+\infty} NZ_m \cdot NX\frac{\sin \pi NX(u - mU)}{\pi NX(u - mU)}.$$

(63)

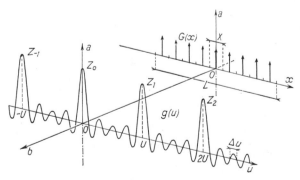

Fig. 131.

This expression would be identical to Eq. (13) if x were changed to u. The expression that is equivalent to Eq. (15) may be directly written as

$$g(u) = N^2 \sum f(mU) \frac{\sin \pi NX(u - mU)}{\pi NX(u - mU)}. \tag{64}$$

The well-known result of this convolution is shown in Fig. 131 for a grating having 9 periods; the function $F(x)$ has been left in its $\delta(x)$-function form to avoid showing the modulation of the Z_m. The main maxima and the N-2 secondary maxima are easily recognized in the figure.

Third Operation

This third and last operation is the most simple: the spectrum—that is, the diffraction pattern $g(u)$—is limited. Its mandatory limitation to the hemisphere of extension is due to the inadequacies of optical systems. Crystals diffract x-rays well for the whole sphere of extension. The poor gratings that we formerly made with collodion from plane Rowland gratings diffracted well, both in transmission and in reflection. Certain opaque gratings traced on plane mirrors were used from normal incidence to almost grazing incidence. It seems reasonable to think that the limitation to the hemisphere of extension of the diffraction patterns of gratings is a disagreeable but unavoidable accident.

How will this limitation manifest itself? The imaging of a grating by means of wavelengths for which the grating is designed is of very limited interest, and modern techniques require electron-microscope images.

The limitation of the spectrum can only force the replacement of the continuous and wavelike $G(x)$ by a $G_0(x)$ broken up into information points, as we have previously discussed. The only difficulty is the accuracy of the periodicities. It seems to us that the distribution of the information points should be periodic.

GRATING DEFECTS

Grating defects have been the nightmare of optical technicians for over a century: the working of metal has long remained less accurate than the working of glass. The rapid development of spectroscopy at the end of the nineteenth century brought the work of some great scientists to bear on the problem: Rowland, Michelson, Millikan, and Wood. Present-day gratings, with the help of new techniques, are masterpieces of precision.

Some photographs of grating spectra for various exposure times are shown in Fig. 132. The grating used was a plane Rowland grating acquired by C. Fabry around 1900, and it was lent to me by H. Buisson in the thirties. Its rectangular area measured 80×60 mm and contained 45,700 lines: the period X was therefore 1.75 μm. It was used in the second-order spectrum, with an excellent achromatic objective having a focal length of 2 m, and an excellent Jobin symmetrical slit adjusted at 10 μm. The shortest exposure is 3 seconds, and the longest is 90 minutes, a ratio of 1 to 1800. After 30 seconds, the strongest spectral line is surrounded by four symmetrical ghosts, which in the longest exposures become a more numerous periodic set. This grating was not one of the best that Fabry had, but it was very useful for making preliminary measurements on line spectra.

The Perfect Grating

This is the grating that every manufacturer would like to make. The number of lines N is given by the counter on the grating engine; the translation of the grating in the engine, $(N - 1)X$, is measured by the dividing mechanism. For a plane grating, the function obtained is best determined from the monochromatic spectrum with a wave length λ incident perpendicular to the grating. The surface of the grating is three-dimensional. For the present, the prudent course would be to determine $F(x)$ from a good electron micrograph of the lines.

The Real Grating

If we want to determine completely the optical properties of the perfect grating, it must be illuminated with a parallel beam of light, and its Fraunhofer diffraction pattern examined as shown in Fig. 75b. The grating is two-dimensional, rectangular and:

1. independent of y over the range ψ of the range function $\psi(x) = 1$ [it yields along v a narrow diffraction pattern $\Psi(v)$, all the more negligible when the focal point of the collimator is replaced by a slit parallel to y, in which case the distribution is integrated on the u-axis];
2. periodic in x, yielding the special dispersion of Eq. (63):

$$g(u) = Nf_0(u) \otimes l(u). \tag{65}$$

The function $G(x)$ may be replaced by its discrete counterpart G_0. In an apparatus for which the variable of interest is λ, the information points will be chosen with a period of $\lambda/2$. The preceding expression becomes

$$g(u) = NZ_m \otimes l(u). \tag{66}$$

Now if a real grating with its defects is put into the parallel beam, the two diffration patterns along u and v are modified. It is to be hoped that the diffraction along v will not be much modified. It is not affected by the irregularities of the lines, but only by the irregularities in reflecting power. Rowland gratings degraded rapidly in this respect, and efforts to regenerate them rarely succeeded. In any case, the integration along y, especially if the collimator has a slit, still concentrates the diffraction along the u-axis.

The information points I' of the imperfect grating coincide with the information points I of the perfect grating. Therefore

$$I' = IJ, \tag{67}$$

where $|J|$ is multiplied by $|I|$ and the phase of J is added to that of I. The relation

$$G'_0 = G_0 J_0, \tag{68}$$

between the discrete functions has the corresponding transforms

$$g'(u) = g(u) \otimes j(u),$$
$$= NZ_m \otimes j(u) \otimes l(u). \tag{69}$$

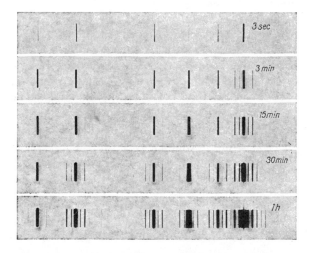

Fig. 132. Ghosts in a 45,700-line planar Rowland grating in an autocollimation mode in second order (photo by P. M. Duffieux and L. Grillet). Beginning with the 3-minute spectrum, a first-order line appears at the left. It is extraneous and is in the UV. The convolution expands with the length of the exposure.

The function $l(u)$ has been placed last because it is the least visible of the three convolution factors.

The ghosts that surround the principal maxima NZ_m of the monochromatic spectrum of the grating are each the spectrum of the error function $J(x)$. They are replicated about each information point NZ_m; their intensities are proportional to those of the information points. Their limits correspond to the resolving power for the wavelength at the information point.

Gratings are mostly used to decompose the spectra of complex radiation patterns. The x- and y-values should be measured in absolute units. The frequencies are then also determined in absolute units. Then for any wavelength, the error function and in consequence its spectrum (the ghosts) have the same width along u. The wavelength only determines where they are located and their extension. This may be verified in Fig. 132.

From the wavelengths of the lines, the number of lines, and L on a photograph such as Fig. 132, it is an easy matter to determine the frequencies of the ghost lines. This technique of determining grating accuracy is very precise. About twenty years ago, we were told by Professor Inglestam that the wearing down of the diamonds used to draw the gratings makes the spectral lines unsymmetrical. The period of the corresponding error function was then L, but it was necessarily an odd function. This was enough to make the fine spectral lines unsymmetrical.

On The Definition of Frequency

When we began our work on the Fourier transform, an impenitent Bourbakiste told us: "Be especially careful not to treat a periodic function over a finite interval like an infinitely periodic function." In 1935, this was a common attitude among mathematicians. We have seen that the Fourier transform more realistically allows an exact definition of the periodicity of a limited array, from both a theoretical and an experimental point of view, as long as the error convolution on the spectrum of $G(x)$ has a negligible effect. Then $j(u)$ is a $\delta(u)$-function, and the function $J(x)$, everywhere equal to one, leaves intact the regular array of $F(x)$. The periodicity is constructed under the condition

$$j(u) = \delta(u),$$

$$J(x) = H(0).$$

This condition may also be expressed in the following way: the information points of the set of N periods are identical with the individual information points of the periods, multiplied by N. The periodicity is attained by replicating one period the required number of times.

Note on the Retinal Image

The retinal image is obviously discrete and certainly unresolved. This is a consequence of the cellular structure of the retina, its heterogeneous dimensions, the signals along the nerves, and the physical mechanism of vision in the brain. This takes us far from the Fourier transform. We abandon this path all the more willingly because here mathematics and biology are in perfect agreement.

10

Transmission of
Spatial Frequencies

INTRODUCTION

Readers who have read the first edition of this book will find a great similarity between this new Chapter 10 and the old Chapter 9. Most of the old figures and sections are left unchanged, but we have removed the lengthy introduction.

It has been out because it is outdated. Spatial frequencies play a role in defining the properties of images and even of objects; in the definition of resolving power and of the relative and absolute limitation of spectra; in the definition of noise properties, and in the parallel with acoustics and electricity. But all that is now part of the standard curriculum in the graduate teaching of optics, and will gradually descend to the more elementary levels of teaching. We shall therefore only summarize at the start what we know about range functions and about the absolute limits of the complex amplitudes of spectra in one, two, and three dimensions.

After these questions have been dealt with, we shall consider the problem of frequency selection or filtering, the uses of which are constantly increasing and becoming more complex. We shall end with a short summary of the work of Abbe, who was the first to perceive the role of spatial frequencies in coherent imaging.

DOMAINS AND RANGE FUNCTIONS

In Fig. 133 are shown along the same axis representations of the domains or ranges of complex amplitude spectra for one, two, and three variables. For a single image there are never more than two independent variables. These domains are absolute maxima for each category of image. Any experimental or theoretical spectrum of a monochromatic image is contained in the domain corresponding to its type and wavelength; it may partially or completely fill the domain.

Let us begin with the case of two variables, which is the most complete example of the Fourier transform.

$\varphi(u, v)$ and $\varphi'(u, v)$

The function φ is the equatorial circle of a sphere having a radius equal to $\mu = 1/\lambda$, with center at O', Figure 133 only shows the hemisphere corresponding to positive w, oriented in the direction of propagation of the light. The probability density is constant over the whole surface of the circle and defines the range function $\varphi(u, v) = 1$. Its Fourier transform has cylindrical symmetry and not spherical symmetry:

$$\Phi(x, y) = \pi\mu^2 \frac{2J_1(2\pi\mu r)}{2\pi\mu r} \quad \text{with} \quad r = +\sqrt{x^2 + y^2}. \tag{1}$$

This is the Airy disk, whose classical diameter is equal to 1.22λ with an accuracy better than one percent. Its exceptional size is due to Lambert's law, which is followed by a light point with the elementary area $dx\,dy$.

The hemisphere of radiation φ', which we associate with the circle of extension φ', is the locus of the vectors with length μ that represent the directions of propagation beyond the xy plane towards infinity, or more exactly (as we explained in Chapter 6 when we discussed Fig. 75b), towards focusing on a spherical surface. Consider a complete sphere of radiation, with homogeneous surface properties for x-rays, similar to Laue's sphere of

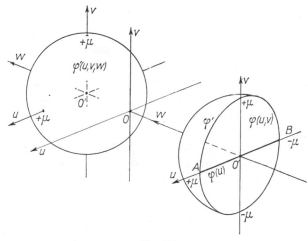

Fig. 133.

extension. For this complete sphere with a uniform probability σ, the range function is

$$\sigma(u, v, w) = 1 \qquad \text{for} \quad u^2 + v^2 + w^2 = \mu^2. \tag{2}$$

But the hemisphere of probability $\varphi'(u, v, w)$ projected on the uv plane must restore the unit density of the circle of extension $\varphi(u, v)$. This leads to

$$\varphi'(u, v, w) = \sigma(u, v, w) \cdot w \qquad \text{for} \quad 0 < w < \mu. \tag{3}$$

The required Fourier transform of $\varphi'(u, v, w)$ must therefore be identical to that of $\varphi(u, v)$, that is, a two-dimensional function able to play the role of internal correlation function for $F(x, y)$:

$$\Phi(x, y, 0) = \int_0^\mu \sigma(u, v, w) \cdot w \, dw \int_\varphi \text{cis}\, 2\pi(ux + vy) \, du \, dv \tag{4}$$

$$= T[\varphi(u, v)] \tag{5}$$

$$= \Phi(x, y). \tag{6}$$

The first integral in Eq. (4) projects $\varphi'(u, v, w)$ on the plane and restores $\varphi(u, v)$; once again we find a classical equation.

We shall return to the delicate case of the sphere of radiation when we consider the case of three variables u, v, w.

$\varphi(u)$

From an experimental point of view, this one-dimensional function does not exist, but it is so representative and so easy to manipulate that we are keeping it. The range is a segment U of the u-axis inside the maximum range $-\mu, +\mu$. In reality the effective domain is never equal to zero along v, but when the xy distributions depend only on x, the value of φ is a constant for a constant v.

The plateau $\varphi(u, v)$ is a rectangle generated by convolving a segment U with another segment V. The internal correlation function is therefore the product of two Fourier transforms of rectangle functions:

$$\Phi(x, y) = UV\left(\frac{\sin 2\pi Ux}{2\pi Ux} \frac{\sin 2\pi Vy}{2\pi Vy}\right). \tag{7}$$

The zeros of this function are lines parallel to the axes:

$$\Phi(x, y) = 0 \qquad \text{for} \quad x = nX = \frac{n}{U}, \quad y = mY = \frac{m}{V}, \qquad (8)$$

where n and m are nonzero integers.

The exactness of these functions, which allow the study of diffraction by rectangular apertures, has been verified experimentally. In general any diffraction function that has rectangular symmetry may be reduced to combinations of functions of one variable.

The Sphere of Extension and the Sphere of Radiation

If a certain confusion between these two spheres may be allowed when only geometrical optics is considered, a clear distinction becomes necessary when the Fourier transform is considered. We have met with Laue's sphere of extension when we discussed the three-dimensional diffraction of x-rays. Consider Figs. 79 and 80 of Chapter 6: In Fig. 79, Laue's sphere of extension is the small sphere centered on O' with a radius equal to $1/\lambda$. It is tangent to the origin of frequencies (u, v, w). Let this spherical volume be called φ, and its range function $\varphi(u, v, w) = 1$; now

$$\varphi(0, 0, 0) = 1. \qquad (9)$$

In consequence, if

$$\Phi(x, y, z) = T[\varphi(u, v, w)], \qquad (10)$$

the Fourier-transform relation implies

$$\int\int\int_{-\infty}^{+\infty} \Phi(x, y, z) \, dx \, dy \, dz = \varphi(0, 0, 0) = 1. \qquad (11)$$

The Fourier transform of the range function of the Laue sphere is normalized. This is also true in Eq. (51) of Chapter 5, which symbolizes our interpretation of Dirichlet's theorem for functions with limited spectra:

$$F(\cdot) \otimes \Phi(\cdot) = F(\cdot). \qquad (5.51)$$

In Fig. 80 of Chapter 6, a similar representation was used for a two-dimensional coherent phenomenon: the oldest example of this, Abbe's oblique illumination, will be discussed later.

The light vectors all have their origin at O', for x-rays as well as for Abbe's illumination, when the phenomenon is the diffraction or the dispersion of a beam of light that is sufficiently collimated to be represented by a simple vector OO'. The locus of all the incident and diffracted directions is the hemisphere above the diameter through O' and marked $-\mu, +\mu$. The plane of the figure is perpendicular to the plane of diffraction and passes through both the vector OO' and the u-axis.

The function $f(u, v, w)$ is represented on the hemisphere by points such as P. It is both a directional function (the vector $O'P$) and a Euclidian function: P is independent of its neighbors. Some of the projections $H(0, w)$ of $f(u, v, w)$ on the uv plane are shown in the figure. The projection $f(u, v)$ is thus enclosed within the projection of the circle of extension having a diameter 2μ.

If the sphere and the hemisphere are translated from O' to O, all trace of the external illumination is lost, as well as any relation to real frequencies. The hemisphere becomes only a hemisphere of radiation for a plane distribution $F(x, y)$: the sphere and the hemisphere of extension are transformed into a sphere and a hemisphere of radiation. The directions are preserved, but the initial frequencies are transposed, yielding in x, y, z space a helical twist or phase shift corresponding to the translation OO'.

At the same time the sphere and the hemisphere both lose their unit value at the origin of frequencies: the range functions are not normalized in a three-dimensional Fourier transform. From the transformation equations, the integrals of the Fourier transforms of the sphere and of the hemisphere are equal to zero, as they are at the origin of frequencies. The value of the Fourier transform of the empty sphere is easy to calculate. From Eq. (63) of Chapter 6,

$$\Phi_0(r) = 4\pi \frac{\sin 2\pi r}{2\pi r} \quad \text{with} \quad r = +\sqrt{x^2 + y^2 + z^2}. \quad (6.63)$$

Integrating on the sphere with radius r and area $\sigma(r)$ yields

$$\Phi_0(r)\sigma(r) = 8\pi r \sin 2\pi r. \quad (12)$$

This is not a convergent function, but the corresponding distribution of energy

$$\left|\Phi_0(r)\right|^2 \sigma(r) = 16\pi \sin^2 2\pi r \quad (13)$$

has a constant average, in accord with the conservation of the radiated energy. As a matter of fact, it is because of this coincidence, which conforms to the definition of the sphere of radiation in directional space, that we have given the sphere its name. We have clearly stated in Chapter 6, under the heading "Infinity," that the meaning of the complex amplitude and its phase changes with the distance r. It is natural that the complex amplitude $\Phi_0(r)$ become undefined when the dispersion of the photons does not allow it to be measured, and when the increased coherence due to the size of the source replaces the initial coherence of the object pupil.

Also in accord with this is the fact that all the images that we can conceive are two-dimensional and indeed plane. In x-ray diffraction by crystals, projections of electron mean densities in a lattice of atoms are drawn on a plane. This is not the result of an observation, but the result of a calculation carried out on the spherical diffraction pattern of the lattice of the crystal. The image is always two-dimensional. One might think that this is due to a human disability that requires images to be projected on a retinal surface. But one should remember that the propagation of light requires one dimension, so that only two are left for observable distributions. Today people are justifiably impressed with holograms that preserve depth. Before holography, stereoscopy was used to give the sensation of depth associated with binocular vision, with all its disabilities. Holograms conform to the mode of acquiring information in our three-dimensional world: two dimensions come from the retina and one comes from movement. We know our surroundings by the touch of the hand, the convergence of the eyes and by walking when the distances are long. Three-dimensional space is heterogeneous, both from a human and from a physical point of view.

TRANSMISSION OF FREQUENCIES

Physicists and technicians think in terms of images, and tend to represent conceptual shapes by means of graphics. The Fourier transform is a general geometry that deals with physically real or imaginary shapes, forms and colors, perspectives and movements. Many diverse kinds of real or imaginary objects are described analytically by means of their spectra: at the Florence ICO meeting in 1954, Professor Inglestam, while discussing these ideas, projected at length on the screen "Duffieux's elephant," shown in Fig. 134, whose integral transform we recently studied in the *Cahiers de Physique*.

There are more serious examples. So-called spectroscopic binary stars are not resolvable and are therefore seen as points. But the harmonic analysis of their spectra yields their period, their relative masses, the differences of their physical states, and so on. The mechanism and the critique of frequency transmission thus plays a considerable role in science and technology.

Let $F(x)$ be a phenomenon that may be represented by a simple periodic function: a helix in complex space, or a sine wave if the function is real. In the language of spectroscopy, this function has only one frequency. After transmission through a physical channel, the original phenomenon or the curve representing it is more or less distorted; the nature of the variable may even have changed. If the variable x is time or luminance, the transmitted function $F'(x')$ may be a curve plotted on a linear scale. The notion of isoplanatism or space invariance may be generalized from optics: the transmission is space-invariant if a translation along x without deformation of the input $F(x)$ results only in a translation of the output $F'(x')$ in its space x'. This is the basis for our Abbe's condition. When the system is space-invariant, the following two statements are true: the transmitted function $F'(x)$ is a function only of x, and $F'(x)$ has the same period as $F(x)$.

The output function $F'(x')$ is the space-invariant or isoplanatic image of $F(x)$. This is the case considered in Abbe's condition; it is also the case in Volterra's closed cycle.

The spectrum of $F'(x)$ may have only one line, like that of $F(x)$, but the scale of the amplitude may be different, and the phase may be changed: this is a phase and amplitude distortion. These defects are usually present in photographic transmission, and we should rejoice that the degree of correction required for holography has encouraged emulsion manufacturers to improve their product.

Let an arbitrary function be built up by adding together functions with different frequencies. If the system is space-invariant for each frequency, the transformation equations may be written

$$F(x) = \int_{-\infty}^{+\infty} f(u)\mathrm{cis}(2\pi ux)\, du, \tag{14a}$$

$$F'(x) = \int_{-\infty}^{+\infty} f'(u)\mathrm{cis}(2\pi ux)\, du. \tag{14b}$$

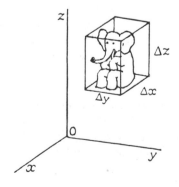

Fig. 134.

When $f'(u)$ and $f(u)$ differ only by a factor that depends only on u and that is independent of the amplitude and the phase of $F(x)$, we may define a transmission factor or transfer function between $f(u)$ and $f'(u)$:

$$f'(u) = f(u)t(u). \tag{15}$$

The convolution of this expression is

$$F'(x) = F(x) \otimes T[t(u)], \tag{16}$$

and the output of the system may be written

$$F'(x) = \int_{-\infty}^{+\infty} f'(x)\mathrm{cis}(2\pi ux)\, du = \int_{-\infty}^{+\infty} f(u)t(u)\mathrm{cis}(2\pi ux)\, du. \tag{17}$$

These are the stringent conditions that a frequency transmission function must satisfy in order to enter into the simple theory of images that we have developed so far. One should not be surprised: it is the nature of mathematical theories to allow discussions and reasonable corrections, rather than to be exact.

We shall limit ourselves, as mathematicians do, to cases where perfection may be assumed, and where imperfections may be measured.

RELAYS AND PUPIL OPTICS

Examples

The transfer of an image through an optical instrument with minimum loss of information and therefore minimum loss of light is an old problem and a delicate one even today. We shall consider only centered systems, which involve only the normal vocabulary and the normal amount of difficulty. For the remainder of this book, we consider only objects and images of one variable. The following are three classical examples.

A Marine Telescope.   In Fig. 135 is shown the optical system of a marine telescope manufactured in England around 1770. This system was studied in a doctoral thesis by Jean Bulabois. The vertical and horizontal distances have been distorted for a clearer representation: the telescope had a maximum length of 1 m and a maximum diameter of 6.5 cm. With the exception of the objective, which was an achromat, all the lenses were thin biconvex lenses with various curvatures. The paths of the two marginal rays are shown in the figure.

The objective *ab* gives a real inverted image in the diaphragm C_1; this image is inverted and slightly magnified by the relay lenses L_1 and L_2. The frontal lens L_1 of the relay and the lens L_3 of the Huyghens ocular play only a secondary role in the transfer of the image from C_1 to C_2: they condense the light about the axis.

The lenses never coincide with the images; old glass was much less transparent than modern glass.

It is remarkable that manufacturers of similar relays, even in modern times, never publish their methods of calculation.

Periscopes. In Fig. 136 is shown a schematic drawing of a submarine or fortress periscope. The figure only shows part of the optical system, consisting of two essentially similar structures: the small and the large relay lenses. The first relay is inside the small tube at the head of the field lens C_1, joined by a cone to the large tube, which contains the remainder, starting from lens L_2.

The periscope contains only three images: the virtual image I_0, given by the input divergent objective lens; the real image I_1, which coincides with the field lens C_1; and I_2, which coincides with the field lens C_2 in the focal plane of the ocular Oc.

The image I_0 is in the front focal plane of lens L_1, and I_2 is in the back focal plane of L_2'; I_1, which coincides with C_1, also coincides with the back focal plane of lens L_1' and the front focal plane of L_2. Apertures D_1 and D_2 are in conjugates planes of field lens C_1. Aperture D_2 and the ocular Oc are in conjugate planes of field lens C_2.

Inside the two relays, apertures D_1 and D_2, located in conjugate planes, are the pupils that limit the extent of the beams of light. On the lenses L the areas occupied by the beams are different for different image points. These four lenses, which make up two compound lenses, transfer the images

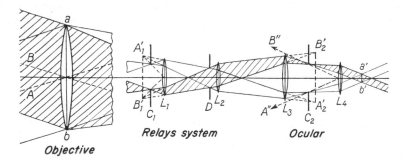

Fig. 135.

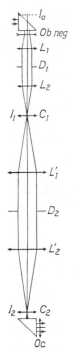

Fig. 136.

through the system. The two field lenses determine where the pupils of the system are.

The small tube and the cone protrude above the enclosure holding the observer; the large tube remains inside. The diameter of the tube in the water is kept small to reduce friction. The cone gives it solid support.

Endoscopes. It could be said of Fig. 137 that it represents not a relay but a convoy. Let S be a luminous object, reduced here to a point; its image is relayed by the odd-numbered lenses $O_1, O_3, O_5, \ldots$, and the even-numbered

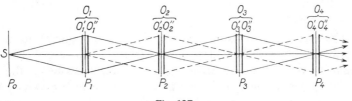

Fig. 137.

lenses have the odd-numbered lenses in their conjugate planes. A light beam from S is shown focusing successively in the principal planes P_1, P_2, P_3, P_4. If a transparency on glass is placed on P_1, lens O_2 and the other even lenses will transmit the images formed in the principal planes of the odd lenses. The role of those field lenses is to conserve the light beams over the field of interest.

Such convoys are used in rigid medical endoscopes and, with modifications, in flexible ones. They are found in all very high-speed electrical particle accelerators such as the Geneva cyclotron. The convergent systems are then electrical or magnetic systems, which have their jobs complicated by the natural tendency of charged-particle beams to diverge.

Let us stay for the moment with incoherent optics, which here corresponds either to self-luminous objects or to objects illuminated by means of large incoherent diffusers.

1. Let such an object be located in plane P_0, at S. The first image is in plane P_2, and succeeding images are in the even planes. The pupils of O_1 and O_3 are in conjugate planes of O_2. If the relay is well constructed, because the apertures of O_1 and O_3 are in conjugate planes, no ray of light is lost. The even lenses are *field lenses* or collectors; they are the *pupil optics* of the system, and it is their function to conduct the light through the system with no losses except those due to absorbtion by the lenses. It is interesting to compare Fig. 136, which has three images between the object and the ocular, with Fig. 137.

In the relay of Fig. 137, each lens O is broken up into two lenses O' and O'' which are symmetrical with respect to the principal plane P. Here this is only a useful fiction, but we have often realized it for monochromatic light by means of two identical plano-convex lenses with their curved surfaces facing each other, as illustrated in Fig. 138. Their apertures cannot much surpass $f/10$. The lenses O' have their front focal planes at the optical center of the previous lens, and the lenses O'' have their back focal planes at the optical center of the next lens. The rays from a point P_{n-1}, which converge to its image at P_{n+1}, are parallel between O'_n and O''_n. On the other hand, if point sources at P_{n-1} are projected by O_n on plane P_{n+1}, then O'_{n+1} projects the pupil of O_n to infinity, and all the diffraction patterns in plane P_{n+1} are identical.

2. The aperture of O_1 is determined by the diaphragm D_1 located at P_1. The diaphragm D_1 and the distance C_1C_2 determine the angular aperture of the beam convergent to the image of S, and in consequence the spectral bandwidth φ for a coherent image and the bandwidth $\varphi \otimes \varphi(-)$ for an incoherent image, where $\varphi(-)$ is the reflection of φ with respect to the

frequency origin. Because D_1 is projected to infinity by O_2', these domains are the same for all points of P_2.

The objective O_2 has a diaphragm D_2 that limits the image in P_2 of an object in P_1. In incoherent optics D_2 is an aperture stop. It limits the number of information points contained in the first image; their period is determined by D_1. This number does not change in subsequent steps through the relay.

Coherent Imaging

In coherent systems, relays and convoys take a completely different form. Let a plane wave from a point source S illuminate a transparency in plane P_1. Let the transparency alter the phase and the amplitude of the incident wave; this is the case if the transparency is a photographic emulsion. Let the spectrum be focused in plane P_2. In theory and perhaps in practice, this spectrum occupies the whole of the hemisphere of radiation. Aperture D_2 lets through only part of the spectrum, depending on the position of S in plane P_0. Aperture D_2 is therefore a spectral bandpass filter, and the image found at plane P_3 depends on which part of the spectrum goes through D_2.

The aperture D_2 played the same role in incoherent imaging, but for each individual point of the object. In the incoherent case, the object must be broken up into its independent points, which may be infinite in number according to Euclidean concepts, or a finite number of information points. The complex Fourier transform applies to each individual point.

Let S be shifted to S' in plane P_0 along a vector h. The plane wave that went from S to P_1 is subjected to a helical twist or phase shift $-2\pi hu$, if h is along the frequency axis u. The image of S in plane P_2 is shifted by the quantity $-h$. The inversion of the image may be considered a consequence of the two successive Fourier transformations in the propagation from the object to the image.

The relay thus rapidly becomes limited; in the most frequently used system, the relay is limited to the observation plane P_3. This configuration, shown in Fig. 138, is known as the double diffraction configuration: the

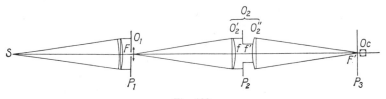

Fig. 138.

diffraction of the coherent object $F(x, y)$ yields the spectrum $f(u, v)$ on P_2, which is reduced to $f'(u, v)$ by the aperture D_2 and the absorption in plane P_2. The image $F'(x, y)$ is observed in plane P_3. The image is sometimes considerably distorted.

The general case is called the double Fourier transform. If $d(u, v)$ is the complex transmittance of D_2,

$$T^*[F(x, y)] = f(u, v), \qquad F'(x, y) = T[d(u, v)f(u, v)]. \quad (18)$$

This analysis is rounded out in the following sections with a series of examples, mostly classical, but which, as we have had occasion to verify, have not yet penetrated the habits of those who manipulate coherent light from lasers. We should not photograph the image at P_3 without looking at the spectrum that has gone through aperture D_2.

Image of a Coherently Illuminated Slit

The analysis of a coherently illuminated slit was published by us in the *Annales de Physique* in 1944. Consider the double diffraction system of Fig. 138, using the notation of Fig. 137. The two lenses O_2 had focal lengths equal to 1840 mm and a diameter equal to 80 mm. Lens O_1 was a thin lens with a power of 0.5 diopter. The objective of the collimator had a focal length of 2500 mm and an aperture of 80 mm. The complex amplitudes in planes P_1, P_2, and P_3 are shown in perspective in Fig. 139. The correct functions $f(u)$ and $F'(x)$ are shown in Fig. 140.

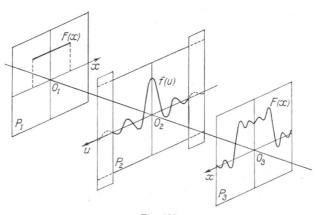

Fig. 139.

In Figs. 140, 141, 142, and 143, $f(u)$ is given as an amplitude but $F'(x)$ is given as an intensity, because the latter is a complex function. Figure 142 is corrected according to the calculations of L. C. Martin (*Theory of the Microscope*).

The image F' in Fig. 143, where only the discontinuities are imaged, is an approximation to strioscopy. The filtering of Fig. 144, where an opaque

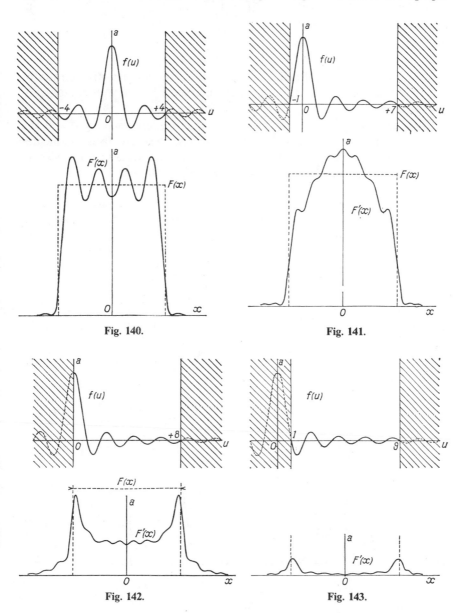

Fig. 140.

Fig. 141.

Fig. 142.

Fig. 143.

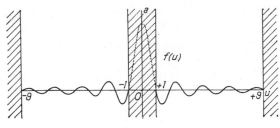

Fig. 144.

band removes the central order of $f(u)$, yields the strange distribution of Fig. 145. The contours of the object are marked by a very fine, high-contrast, dark fringe.

If the spectrum and the central obstruction are left unchanged and the number of fringes is increased to about thirty, the dark fringe coincides exactly with the edge of the slit. This is the most accurate way to measure the optical width of a slit. It is illustrated in Fig. 146.

Note. For the sake of comparison, Fig. 147 shows the intensity distribution in the image of the same slit used in the same double diffraction configuration, but using incoherent light.

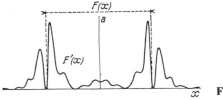

Fig. 145.

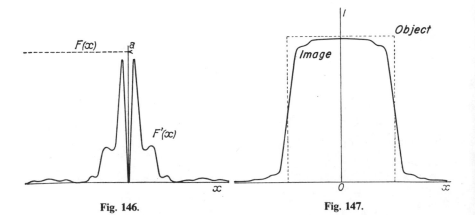

Fig. 146.   **Fig. 147.**

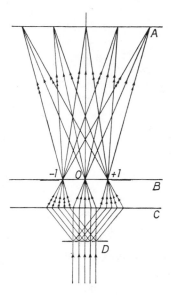

Fig. 148. Images in a microscope in coherent illumination: Abbe's experiment. After Wilhelm H. Westphal (1952); reproduced in the account of Carl Zeiss and Abbe published at Jena for the 150th anniversary of Carl Zeiss. *A*, image; *B*, image focal plane; *C*, principal plane of the objective; *D*, object (grating on glass).

Abbe's Experiment

The paths of the rays in a microscope are shown in Fig. 148. An almost collimated beam from the light source directly illuminates the object without using a condenser lens. The object is a grating of parallel lines engraved on glass. Note that the object that Abbe chose was analogous to the one chosen by Foucault, but adapted to the microscope. When the ocular is removed, the eye can see the spectral lines of the grating in the back focal plane of the objective lens. The extent of this spectrum is reduced by means of an

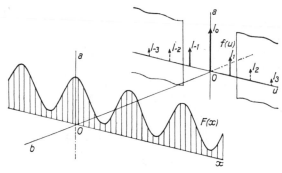

Fig. 149.

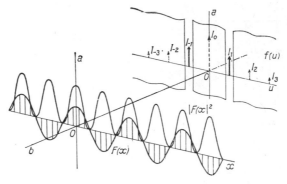

Fig. 150.

absorbing screen. In this plane B, Abbe placed screens that allowed one or more spectral lines through. For instance, in Fig. 149, only the three central orders -1, 0, and $+1$ are allowed to pass.

Figure 149 also shows the Fourier transform corresponding to the central order (a constant) and the two lateral orders (a cosine function). The object being real and even, so is its Fourier transform. In the plane A, the image has the same frequency as the object D.

In Fig. 150 the central order I_0 is removed, leaving only the two lateral orders I_{-1} and I_{+1}, whose superposition yield the final image $\cos 2\pi Xu$, shown in the figure with vertical hatching. But the eye sees only the energy distribution, which has *twice the frequency* of the object D.

In Fig. 151, only the order I_{+1} is allowed to pass. The complex amplitude of $F(x)$ may be represented as a helix. But the eye sees only the intensity $I|F(x)|^2$, which is constant: the structure of the object has completely

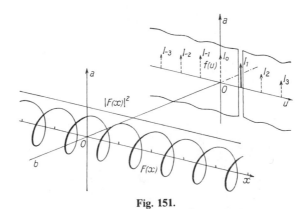

Fig. 151.

disappeared. The result would be the same if any other line were isolated by itself in plane *B*.

Missing from this collection of figures are those for one of Abbe's most significant experiments, where the aperture in plane *B* is left just wide enough to let two successive lines through. If *U* is the period of the lines, the width of the slit is between *U* and *2U*. Depending on the position of the slit, either one or two lines come through the slit.

When only one line goes through, the image has no structure. When two lines go through, the image is periodical and its intensity distribution has a period equal to the period of the grating.

Oblique Illumination

From the preceding experiments, it may be concluded that in order to observe a high-frequency structure, one high-frequency line must go through the system in addition to the central order I_0. In Fig. 152 the screen in plane *B* allows the two extremities I_0 and I_{+6} of the spatial frequency range to go through. The image effectively has a frequency equal to $6U$. Abbe concluded from this that the resolving power of the microscope was multiplied by two.

Let us try to visualize the movements of the spectra in plane *B* for a two-dimensional object. Consider an object with a double periodicity: *U* along the *u*-axis, and *V* along the *v*-axis. The spectrum is shown in Fig. 153.

Let plane *B* contain a circular aperture with a diameter $6U$ whose center *O* is on the optical axis. The orders I_0' and I_{+6}' are at the two extremities of the diameter of the circle *C*. Abbe rotated the plane of obliquity about the axis. Then the conditions of x-ray diffraction and of Fig. 79 of Chapter 6 are approached. Let the object be very thin, so that it may be approximated

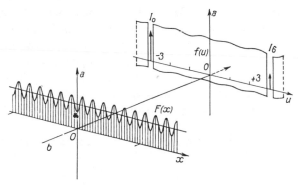

Fig. 152.

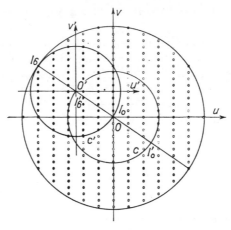

Fig. 153.

by a plane. It remains stationary, and its theoretical spectrum is that of plane B with I_0 on the optic axis. In oblique illumination, the circle is shifted to C' and its center is shifted to O'. The small dark circles represent the object, whereas the small open circles fill the area of the circle of extension C'.

The translation OO' only causes a helical twist or phase shift in the image. The modulus is not changed, and the eye sees no change in the image.

For the diffraction of x-rays, we assumed that outside the atomic nuclei, the index of refraction was very close to one. For the wavelengths of interest in the microscope, the index is not equal to one and it changes over the area of the object even in the thinner samples; these variations are useful in phase-contrast microscopy. There is in general no form function that is independent of the angle of incidence. Oblique incidence will therefore only serve to determine if any particular frequency is outside the circle C. That can be quite useful. In his last experiments, Abbe remarked that the siliceous shell of a certain diatom had along its edge a periodic pattern with the highest frequency that he was able to see. Von Ardenne and his team examined it in their first tests with their electron microscope; its photograph is in the first collection of photographs that they published. Abbe had seen well.

I point this out as an example of the admirable continuity of scientific endeavor.

Index

Abbe's condition, 87, 96, 100, 104, 182
Abbe's experiment, 191
Abbe's oblique illumination, 179
Additivity, 27
Airy pattern, 109
Alternating functions, 9
Amplitude spectrum:
  grating, 122
  slit, 121
Angular frequency, 10
Apodizing pupils, 150
Autocorrelation function, 114

Bandlimited functions, 80
Bessel functions, 49
Bounded sets, 156

Change of origin, 18
Circle of extension, 97, 106
Circular pupil, 145, 163
Coherent illumination, 157
Coherent imaging, 187
Commutativity, 61
Complex functions, 25
Complex integrals, 26, 116
Complex representation, 23
Complex series, 26
Complex transform, 35
Conjugate, 24
Conservation of energy, 116
Convolution, 56, 60, 70ff, 87, 116, 119
Convolutions, multiple, 62, 83
Correlation, 112
  internal, 108
Correlation equation, 125

Delta function, 65, 165
Descartes' law, 96
Diffraction:
  Fraunhofer, 86
  inverse, 91
  plane wave, 88
  X-ray, 101
Dirac, 65
Dirichlet's theorem, 70, 75
  converse, 152
Discontinuities, 68
Discrete functions, 151
Discrete pupils, 161
Distributions, two-dimensional, 104
Domains, 176
Duffieux's elephant, 181

Endoscopes, 185
Eppler's faltung, 78
Even functions, 7
Ewald, 102
Exponential spectra, 43

Faltung, 78
Fermat's principle, 86
Field lens, 186
Finite grating, 167
Fourier integral, 1ff, 13
Fourier series, 1ff, 115
Fourier transform, 20ff, 67, 95
Fragmented function, 5
Fraunhofer diffraction, 86
Frequency, 10, 174
Frequency space, 20ff
Frequency transmission, 181
Fresnel's mirrors, 131

Gaussian function, 46, 78
Ghosts, 173
Graphical representation, 12, 14, 20, 123
Graphical spectrum, 169
Grating, 37, 165
  amplitude spectrum, 122
  finite, 167
  perfect, 172
  real, 172
Grating defects, 171

Half-space transformation, 32
Hermite functions, 48
Huyghen's function, 102
Huyghen's principle, 87

Illumination, oblique, 193
Image, retinal, 175
Imaging, coherent, 187
Impulse response, 73
Incoherent illumination, 156
Incoherent imaging, 87
Infinity, 92
Information points, 158
Integrable points, 66
Integrals, 20, 55
Integration, images, 59ff
Interference, 128
Interference fringes, 128
Internal correlation functions, 108
Inverse diffraction, 91
Isolated limited function, 5
Isoplanatic images, 56
Isoplanatism, 100
Isosceles triangle, 42

Jumps, 54

Lambert's law, 108, 111
Laue, 102
Laue-Epler sphere, 87
Laue sphere, 179
Limited functions:
  image, 62
  isolated, 5, 15
Limited isolated function, 15
Linear ramp, 39

Martin, 189
Multiple convolutions, 62, 83

Normalized form, 6

Oblique illumination, 193
Odd function, 8
One-dimensional image, 58

Parity, 28, 117
Parseval's theorem, 50, 60, 112
Perfect grating, 172
Periodic functions, 4
Periscopes, 184
Pinhole camera, 70
Plancheral's theorem, 112, 114ff
Plane wave diffraction, 88
Point source, 65
Propagation, 97, 104
Pupil optics, 183
Pupils:
  apodizing, 150
  circular, 145, 163
  discrete, 161
  slit, 138
  square, 143
  stigmatic, 128, 137
  symmetrical, 134
  two-dimensional, 141
  two variable, 89

Quadratic spectra, 119
Quadratic spectrum, 125

Ramp, linear, 39
Range functions, 108, 176
Real grating, 172
Reciprocity, 91
Rectangle function, 36, 77
Refractive index, 95
Relays, 183
Representation:
  complex, 23
  series, 2
Resolution, 155
Resonance curve, 73
Retinal image, 175

Scanning slit, 71

Series, 21, 63
Shape function, 101
Singular functions, 51
Singular points, 54
Slit pupil, 138
Slit:
    amplitude spectrum, 121
    scanning, 71
Spatial frequency,
    176
Spectra, 12
    exponential, 43
    quadratic, 119
Sphere of extension, 179
Sphere of radiation, 106,
    179
Square pupil, 143
Stigmatic images, 57
Stigmatic pupils, 128,
    137
Symmetrical pupils, 134

Symmetry, 7, 28

Telescope, marine, 183
Time techniques, 78
Transform, complex, 35
Transmission, 176
Transmittance, plane diaphragm,
    40

Uniqueness, 19

Volterra's closed cycle,
    182
Volume of integration, 119
von Ardenne, 194

Weighting functions, 75,
    77
Wood, 132

X-ray diffraction, 101